FOOD, CLIMATE, and CARBON DIOXIDE

The Global Environment and World Food Production

Sylvan H. Wittwer

LEWIS PUBLISHERS

Boca Raton New York London Tokyo

Library of Congress Cataloging-in-Publication Data

Wittwer, S. H. (Sylvan Harold), 1917–
Food, climate, and carbon dioxide: the global environment and world food production / Sylvan H. Wittwer.
p. cm.
Includes bibliographical references (p.) and index.
ISBN 0-87371-796-1
1. Crops and climate. 2. Climatic changes. 3. Crops—Effect of atmospheric carbon dioxide on. 4. Global warming. I. Title.
S600.7.C54W57 1995
338.1′4—dc20 94-47271
CIP

This book contains information obtained from authentic and highly regarded sources. Reprinted material is quoted with permission, and sources are indicated. A wide variety of references are listed. Reasonable efforts have been made to publish reliable data and information, but the author and the publisher cannot assume responsibility for the validity of all materials or for the consequences of their use.

Direct all inquiries to CRC Press, Inc., 2000 Corporate Blvd., N.W., Boca Raton, Florida 33431.

Lewis Publishers is an imprint of CRC Press

International Standard Book Number 0-87371-796-1
Library of Congress Card Number 94-47271
Printed in the United States of America 1 2 3 4 5 6 7 8 9 0
Printed on acid-free paper

DEDICATION

The principal characteristic of climate is variability.

Sylvan H. Wittwer is Director Emeritus of the Michigan State University Agricultural Experiment Station (East Lansing), and Professor of Horticulture. He received his B.S. degree in 1939 from Utah State University (Logan), and his Ph.D. from the University of Missouri at Columbia in 1943 in the field of horticulture.

Recognized as a world authority on greenhouse crop production and protected cultivation, he has published over 750 research papers and scientific reports, and conducted pioneering work on atmospheric carbon dioxide enrichment for enhancing the production of food crops. He was chairman of the Board on Agriculture of the National Research Council 1973–1977; a member of the Climate Research Board of the NRC 1979–1981, and organizer and co-chairman of the International Conference on Rising Atmospheric Carbon Dioxide and its Effects on Plants, Athens, GA, May 23–29, 1982. He cochaired and organized the International Conference on Crop Productivity-Research Imperatives, Boyne Highland, MI, October 1975 and in October 1985.

Serving as a consultant for all International Agricultural Research Centers, all U.S. federal agencies, the Ford and Rockefeller Foundations of the United Nation's Development Program and World Bank relating to food production, he has written extensively on technical developments in agriculture, food security, and prospects for the 21st century.

He was Director of the Michigan State University Agricultural Experiment Station for 20 years until his retirement in 1983; was one of only two foreign members in the United States ever elected to the V. I. Lenin All-Union Agricultural Academy of Sciences. He is first author and editor of *Feeding a Billion* which is a major volume, co-authored with three Chinese scholars on frontiers in Chinese agriculture.

ACKNOWLEDGMENTS

This book has an interesting genesis. I greatly reduced my involvement in the climate change, carbon dioxide, and global food production discussion over a decade ago.

Encouraged by Robert "Skip" DeWall of Lewis Publishers, Adam Meyerson, editor of *Policy Review* and others, Paul Waggoner, Norman Rosenberg, Norton Strommen, Hartwell Allen, and Hugo Rogers, I decided to take on the task, not fully realizing its enormity.

I extend thanks to the above and to many others including Sherwood Idso, Bruce Kimball, Roger Dahlman, Joe Sullivan, Edwin Fiscus, Bert Drake, Pat Michaels, Martin Parry, Joe Ritchie, Hugh Ellsaesser, Robert Balling, Bill Kellogg, Tom Karl, Stanley Changnon, and Sangke Hahn for providing timely updates of books, articles, manuscripts, reprints, and preprints.

Special thanks are extended to Norton Strommen, Howard Camden, Hugo Rogers, Hartwell Allen, Sherwood Idso, and Paul Waggoner for reading drafts of the book and offering helpful suggestions.

Finally, I am grateful to Nancy Washburne and LaRee Farrar for editorial assistance and for reviewing the entire manuscript. Most of all to Maurine, my devoted loving companion and wife of 57 years who transcribed countless draft copies, through to the final preparation of the manuscript.

Contents

1

INTRODUCTION

1.1 SCOPE OF THE VOLUME

The planet earth is a wonder to behold. It appears from satellite imagery as a blue gem floating in the sea of the universe. There is a confluence of climate, sunlight, temperature, precipitation, atmosphere, land, water, and biota that admirably provides for an abundant life for over 5.7 billion human inhabitants. Into the earth's climate system has come, what some perceive, as an anthropomorphically-induced environmental hazard to all life-supporting systems. It has become known as global warming or the "greenhouse effect", a term known by almost everyone, and with many negative implications.

Until the mid-1970s, global warming was not prominent on any environmental or research agendas. It was not a consideration in the World Food and Nutrition Study (National Academy of Sciences, 1977). There was, however, a growing concern among a small group of scientists that the rising level of atmospheric carbon dioxide (CO_2) might lead to an increase in global air temperatures and shifts in precipitation patterns. Such changes might impact agricultural productivity, the output of rangelands, forests and fisheries, and global food security.

The CO_2 in the atmosphere has risen in concentration from a preindustrial level of about 280 ppm to a current level of 360 ppm, a 30% increase. This may change the radiative properties of the earth, and then the climate, by altering the atmosphere's ability to transmit shortwave solar and longwave radiation emitted by the earth and atmosphere. The process of warming the lower layers of the earth's atmosphere caused by the infrared absorptive behavior of CO_2 and other radiatively active trace gases has become known as the "greenhouse effect". The extent and timing of the projected warming, or whether it even exists, are still heatedly debated (Balling, 1992; 1993; Brooks, 1989; Hansen et al., 1993; Karl, 1993; Karl et al., 1991; Lindzen, 1993; Michaels, 1992; Nordhaus, 1994).

Nevertheless, there are some widely-accepted facts. There is a documented increase in atmospheric CO_2. The Mauna Loa record shows a 12% increase in the mean annual concentration, from 316 ppm by volume of dry air in 1959 to 354 ppm in 1990. The current rate of increase is about 0.5% or 1.6 ppm per annum. Carbon dioxide source-sink models predict that the current level of atmospheric CO_2 will be doubled by the latter part of the 21st century.

Second, the increase is truly global. This illustrates the effectiveness of the earth's atmosphere in dispersing emissions from whatever the source, be it natural or man-made.

Another well-known fact from the Mauna Loa record is the annual oscillation of earth's atmospheric CO_2. Although the average level of CO_2 during the year is rising, there is an annual fluctuation. The CO_2 level begins to fall in the spring and continues to fall through the summer months, as CO_2 is removed and appears to be sequestered by the vegetative cover of the northern hemisphere (Gates, 1985). In the late autumn, there is a resurgence of CO_2 into the atmosphere, resulting in new heights by midwinter. With the amplitude increasing by about 0.5% each year, it appears that the concentration of the earth's biomass is either increasing or is steady. It is not decreasing.

There are two ongoing global experiments the outcomes of which we do not know. First, is the greenhouse (heating) effect of increasing atmospheric CO_2 and other radiatively active trace gases, which, according to model predictions, is expected to lead to a warming of the planet ranging from 3.4° to 9.4°F (1.9° to 5.2°C) above preindustrial levels when there is a doubling of the present CO_2 level. Thus, in our span of human history according to current climate modeling, the earth's climate is expected to change more rapidly than it has over any other comparable period of recorded history. This amount of heating, even at the lowest level, is being vigorously challenged by some (Balling, 1992; Idso, 1989a; Seitz et al., 1989). Second, is the stimulative effect of atmospheric CO_2 enrichment on plant growth and development, leading to an increase in global vegetative or biological productivity, and an increase in water use efficiency.

Elevated levels of atmospheric CO_2 tend to partially close stomates of plant leaves, thereby decreasing transpiration per unit leaf area and raising leaf temperatures. This direct effect of CO_2 on plant temperature, coupled with potential air temperature changes (increases) and increases in water use efficiency from partial stomatal closures, will cause changes in plant development, which will, in turn, be superimposed on the direct effects of CO_2 in plant development. Thus, what are the effects of temperature variability (increases?) for CO_2 productivity relationships, and will increasing temperatures augment or reduce the stimulative effects of atmospheric CO_2 on plant growth and development? There could be benefits to the biosphere and society in general from higher levels of CO_2 (Wittwer, 1992).

In 1985, I stated the following concerning atmospheric CO_2: *"Seldom has CO_2 even deserved mention in monographs or books on plant nutrition. Yet, it gives the most remarkable response of all in plant bulk, is usually in short supply, and is nearly always limiting for photosynthesis"* (Wittwer, 1985). Even now, although hundreds of experiments have been conducted in greenhouses, growth chambers, and open fields, the direct positive effects on photosynthesis, photorespiration, dark CO_2 fixation, water use efficiency, consumptive water use, and protection of plants from pollutants have received only cursory mention in many notable and highly visible publications and documentaries on global warming. Almost invariably, the positive, beneficial and directly measurable effects of elevated levels of atmospheric CO_2 are glossed over lightly or submerged in a series of catastrophic issues. These are associated with a widely promulgated global warming, which has been greatly exaggerated, for which the real world climate records have given no confirmation, and which may not even exist.

It is also a sad commentary on the apparent widespread ignorance, even in some segments of the scientific community, about our dependence on plants and photosynthesis for food, oxygen, beverages, the ozone layer, clothing, fossil fuels, biomass fuels, wood, paper, medicines, plastics, perfumes, photographic film, ink, paint, flowers, lawns, shade trees, forests, animal life, and most of our housing. The potential for enhancing the output of all the above products through atmospheric CO_2 enhancement need careful assessment.

During the past decade many atmospheric scientists, and others, with encouragement from science editors, have gone out of their way to avoid emphasizing or even mentioning such benefits. They, with a consenting public, have dwelt on apocalyptic hazards of global warming with speculations of disasters destructive for world agricultural production (Brooks, 1989). They include lack of food security; weeds, insects, and diseases overpowering food and fiber crops; shifts in geographic precipitation patterns that will make deserts out of current food-producing areas; rising sea levels and melting, retreating, and shifting glaciers, flooding picturesque lakeside villages and turning alpine resorts into no-ski areas; sinking of agriculturally productive coastal areas and destruction of arctic tundras and other fragile ecosystems; a shifting of species distributions in forests and the death of trees (Houghton and Woodwell, 1989); and the exacting of a heavy toll on the earth's biodiversity (Dobson, 1992). Indeed, Vice President Albert Gore in his best selling book, *Earth in the Balance,* has solemnly declared that "the process of filling the atmosphere with CO_2 and other pollutants—is a willful expansion of our dysfunctional civilization into vulnerable parts of the natural world" (Gore, 1992). At the conclusion of a Hearing on Global Change Research: Global Warming and the Biosphere of the Senate Committee on Commerce, Science, and Transportation held April 9, 1992, Senator Gore consistently ignored scientific results on the direct effects of CO_2, intent on finding exceptions to cast doubt on any beneficial effects, because of mineral nutrient deficiencies, especially with native species. He concluded that catastrophic climate change will overwhelm the "tiny direct effect of CO_2", and the main consequence will be severe environmental impact. Not only are many of the environmental scientists, the press, science editors, and politicians guilty of overly pessimistic reports, but even segments of the agricultural and forestry community are at fault, as reflected in statements such as the following: "The detrimental implications of global warming for agriculture and forestry are far-reaching and include such phenomenon as increasing frequency of droughts, increasing variation in extreme weather events, and increases in average temperatures, which will push crops and forests to more northerly latitudes" (Cowling, 1987). Pimentel (1991) states that "The projected global warming is expected to increase temperatures generally and reduce rainfall and water availability to crops, especially in the temperate regions." Particularly misleading is a statement by Young (1993) of the Worldwatch Institute, Paper No. 115, entitled, *Global Network—Computers in a Sustainable Society*. "This computer-based modeling of the atmosphere has produced a remarkable consensus among climatologists about the likelihood, and potential scope of global warming. If—as those scientists now predict—"the earth's atmosphere warms by several degrees within the span of a few decades, there will be enormous impacts on the environment and the global economy. Among the predicted impacts are a rise in sea levels that would threaten coastal populations, and shifts in temperature and rainfall patterns that could devastate agricultural areas and force migration or extinction of many of the earth's species." According to Bach (1994), "Global climatic change poses one of the greatest threats to society—The danger is real that our climate and environmental systems will be altered irreversibly if the present misguided policy continues in such areas as economy, energy, agriculture, population and environment. The very existence of all of mankind is a stake, when nature is no longer capable of buffering the adverse effects, both of a reckless exploitation of the natural resources and the ongoing environmental destruction." In summary, the overall consequences of a global warming have been declared so threatening that many scientists, concerned citizens, and political leaders have been urging immediate action to halt the warming (Houghton and Woodwell, 1989). In this vein, one boldly declares, "to give you the real story" (Nance, 1991). We have been in the thrall of model predictions and what the media have dished up for us (Benarde, 1992).

While future CO_2-induced shifts in climate are highly debatable, increased atmospheric CO_2 levels will directly affect vegetation in both native and agroecosystems (Idso, 1989a; Rosenberg, 1982; Rosenberg et al., 1989, 1990; Smit et al., 1988; Strommen, 1992). Evidence will be given, with the support of scientists having agricultural backgrounds, that the benefits from rising levels of atmospheric CO_2 on agricultural productivity will likely outweigh the attendant presumed hazards of a still predictive global warming. It is recognized, however, that the effects on agriculture could range from the potentially beneficial to the seriously disruptive. Further, that even with projected global warming, farmers and the agricultural research and educational establishments in both developed, and agricultural developing, countries can cope and adapt. Farmers have always dealt with weather and climate limitations. Interannual variations and disturbances in climate and weather, with which they yearly cope, are of a magnitude approaching those projected for global warming, where atmospheric CO_2 levels may double. Included are the climate and weather hazards often encountered by a single crop during one growing season. *Climate variability has a greater impact on agricultural productivity than does climate change* (Abelson, 1992, Kane et al., 1992; Katz and Brown, 1992; Wittwer, 1980).

This book will be global in its scope. It will relate to the rising level of atmospheric CO_2 and other atmospheric gases that may alter the radiant heat emissions or radiative forcing of the earth as well as the predictive warming trend. The effects on both the rising level of atmospheric CO_2 and any warming trends will be considered globally as well as their combined effects on the major food crops of the earth, the fiber and oil crops, the rangelands, the forest trees, and food animals. We will address the complex problem of factoring in the direct biological effects of a rising level of atmospheric CO_2 with the presumed climate changes resulting in increases in atmospheric temperatures and precipitation patterns on agriculturally important regions of the world. This report will not cover, in any detail, the response of natural ecosystems (tundra, rain forests, etc.), or species allocation or distribution or survival therein from elevated levels of atmospheric CO_2 with any accompanying global warming. Projections concerning a rise in sea levels and the innundation of agricultural land and coastal areas, and disruption of fisheries, with impacts on food production, are still highly speculative, and will receive minimal note. Projections relating to depleting of the earth's stratospheric ozone levels in the northern and southern hemispheres will be given some coverage, along with the predictive effects of reducing crop yields.

It is further recognized that predictive climate modeling, especially global circulation models, have severe limitations, and little credence is given them. Successful crop modeling with multiple factor inputs relating to productivity is in its infancy. Crop modeling even for single factor inputs, such as heat sums for predicting maturity and scheduling times of harvests and times for processing in sweet corn and peas, is of limited value. General circulation predictive climate modeling is one dimension. Modeling the productivity responses of plants and ecosystems is a much more formidable task. It requires a predictive understanding in the disciplines of physiology, genetics, organismic biology, population dynamics, community dynamics, soil physics and chemistry, and systems ecology. To add to this complexity, the direct benefits and possible hazards to plants of a rising level of atmospheric CO_2 as well as the development of the methodology to determine this effectively remains a still more formidable task in global and crop productivity research (U.S. Department of Energy, 1992).

Putting disciplinary research together is the most challenging task of all. The relationship between climate and water alone cuts across the boundaries of classic disciplines of science. Although separate research by different disciplines may not be wasted, collaboration by scientists of diverse disciplines is the ultimate challenge. This is the requirement to resolve the many unknowns relating to climate and crop modeling, with the added dimensions of a rising level of atmospheric CO_2 (Experiment Station Committee on Policy, 1994).

Rather than an imminent environmental disaster, we will show that the rising level of atmospheric CO_2 represents a resource for agriculture, food production, and forestry not pollutant. The direct (nonclimatic) effects of CO_2 by themselves will be beneficial to agriculture and forestry. It has not yet been determined conclusively whether, on average, global agricultural potential will increase or decrease, but the evidence is strong that it will increase. A review of the contents of this volume, and time, will help resolve that issue.

There is reason for both hope and caution in the future. Agriculture is not as vulnerable today as it has been in the past, to unfavorable climatic conditions. Controlled environment agriculture through crop irrigation (now 17% of all arable land), windbreaks, and protective soil and plant covers (several million hectares) together with new potentials in genetic engineering for frost, heat, and pest protection offer increasing hope for the future. Whether global warming could alter this vulnerability—for better or worse—depends on where, and the ways, climate changes. Soil, water, and genetics could also be decidedly affected by climate change. It is possible that the harm induced by significant greenhouse warming (if it occurs) might outweigh the benefits of a CO_2 fertilization effect, and that of a warmer climate and longer frost-free growing seasons (if they occur). From the Report of the DOE Multi–Laboratory (1990) comes this cautionary note: "Projections of possible climate changes are of sufficient magnitude and reliability to justify consideration of potential impacts on health, water supplies, agriculture, ecosystems, coastal areas and other resources. Initial investigations into the sensitivities and vulnerabilities of these resources to composition and changes of climate, especially storms and other low frequency, but hazardous events, suggest that many, but not all resource impacts will be negative, particularly if the rate of accelerations above its recent rate of about 0.5°C per century."

From the Department of Energy TRO54 report of 1992, entitled "Modeling the Response of Plants and Ecosystems to Elevated CO_2 and Climate Change," comes this report: "In essence, increasing atmospheric CO_2 is an earth-scale experiment (Ramanathan 1988) since, at the present time, it is impossible to experimentally determine the long-term outcome of elevated CO_2 on ecosystems. To predict the outcome of this experiment is a daunting challenge, particularly across the various levels of organizations in the biogeosphere."

Because of the uncertainties of the future, and difficulties involved in reducing the output of other so-called greenhouse gases (methane, nitrous oxide) and the relatively small role of chlorofluorocarbons in radiative atmospheric forcing (Council for Agricultural Science and Technology, 1992; Rosenberg, 1988), efforts should be made to reduce the output of CO_2 into the atmosphere through the conservation of fossil fuels and tropical forests. This will be a play-safe, no-regret strategy with benefits aside from the possible alleviation of any greenhouse effect.

This book will address the following issues: Climate is significant in the distribution, history, and productivity of agriculture and food security. There is the realization that climate should be viewed both as a resource to be managed wisely and a hazard to be dealt with. There is the a rising level of atmospheric CO_2 and other radiative gases and increases will continue. Climate and weather resources that impact agricultural productivity will be identified and the extent to which they are, and will be, modified.

This book takes a two-pronged approach. If climate does not change, or if the change is minimal, as one group of scientists believe, the increasing atmospheric CO_2 concentration will significantly affect the growth of plants and increase agricultural and forestry productivity and will likely extend the boundaries for food production. If climate does change and some warming occurs, there will be both global and regional effects in crop and livestock distribution and of forest tree species. Significant interactions will likely occur between the CO_2 enhancements and the climate variables on plant growth. These will, insofar as possible, be identified. Global warming will not only affect crop plants and their distribution responses to elevated levels of atmospheric CO_2, but will also affect the weeds and other pests with which they compete and the competitive advantages among species in the wild. If the rising level of atmospheric CO_2 is accompanied by global warming, the physical, biological, and chemical properties of soils will change. Soil organic matter content will likely be somewhat depleted. Water resources will be most critical, if hotter and drier conditions prevail, although partial or total compensation may be induced by greater efficiency in water utilization from plants in a CO_2-enriched atmosphere.

This issue will be between first, the presumed and widely publicized impending calamitous events that may climatically overtake world agriculture as a result of the rising atmospheric levels of carbon dioxide and other so-called "greenhouse gases" and second, the ability of farmers and foresters and the agricultural research community to autonomously and directionally adapt. Seasonal and interannual variabilities in climate have always made agriculture uncertain.

The prospects of climate changes from increasing levels of atmospheric carbon dioxide do not frighten many agriculturists, farmers, and foresters. We will suggest that agriculture and its national and international research establishments can cope with, and perhaps even improve during climatic change. History has demonstrated agricultural resilience to change. The rapidity of agricultural change is also well documented. The adaptability of agriculture can be measured historically. Reassurance of resiliency of agriculture and forestry to projected climate changes can be drawn from the present geographic range of major food crops, forest trees, and climate analogs of the past. The surety of climate variability and possible climate change, CO_2 induced or otherwise, should force major research initiatives using genetic resources and management practices to alleviate climatic stresses and improve renewable resource productivity. Genetic differences in response to high levels of CO_2 should be noted and agricultural diversity encouraged. Tuber and root crops, tropical crops, fiber crops, and those having biological nitrogen fixation relationships should be included in further assessments of direct CO_2 effects as well as cereal grains, legumes, and forest, grassland, and range species. Suggestions concerning directions for future agricultural research will be given, accompanied by appropriate government policy strategies.

The future of and the impacts on global agriculture are uncertain. The steady enrichment of the atmosphere with CO_2 and other greenhouse gases could make some global warming likely. From the output of models, both general circulation and crop, we know neither the magnitude nor the rapidity of change nor regional differences. The important issue is whether food security will be endangered; whether crops, livestock, fisheries, and forests will be seriously dislocated or distorted; and to what extent farmers and forestry can autonomously or directionally adapt to a degree or two of warming. There are important analogs of past harmful climate changes and agricultural adaptations through research and cultural practices that provide good examples of what can be done in the future to cope with global warming, if it occurs, as well as interannual climate variations and their impact on global food production. These will be reviewed.

The central focus of this review is that adaptations using modern tools plus new farm technologies, an improved photosynthetic capacity, and water use efficiency of plants from more CO_2, coupled with favorable government programs, will ensure the continued integrity of agriculture and forestry irrespective of any variability in climate in meeting the food and fiber needs of an expanding population during the 21st century and beyond. Scientists cannot now predict, irrespective of all their modeling, the extent of global warming and the effects of it on plants, or accurately reconstruct past climates from pollen or fossil records without learning how to take into account the direct effect of rising levels of atmospheric CO_2 on plants and their evolution. Neither can they predict the ultimate fate of the biosphere (Polley et al., 1993). Thus far, food and fiber production have not only survived by improved with the vicissitudes of climate, politics, and changing increasing demands for improved diets.

1.2 ROLE OF CLIMATE IN AGRICULTURE AND HUMAN PROGRESS

From the beginning, agriculture has been, and remains, the foundation for human civilization. It is the business of agriculture to collect and store solar energy as food energy in plant and animal products. The capacity of the earth's renewable energy resources is enormous. Approximately 90 to 95% of the dry weight of plants is derived from the photosynthetically fixed atmospheric CO_2. It is estimated that over 100 lb of cellulose per day is produced for each of the world's current 5.7 billion people (Mandels et al., 1973). Cellulose is the world's most abundant organic compound and is annually replenished. These outputs of food and fiber sustain the world's population. Historically, agriculture was the first science and throughout most of history has been and remains the most advanced and biggest science of all. Agriculture is the science that makes human life possible. It may be that before this century is over, the success or failure of science as a whole will be judged by the success or failure of agriculture (Mayer and Mayer, 1974). No scientist performs a greater act of faith in the predictability of the operation of natural laws, particularly climate and weather, than a farmer who plants seed and later plows a part of the year's biomass harvest back into the earth. *Food production remains our most important renewable resource.*

Climate and the weather can be proclaimed as the most determinant factors, both for plant growth and for crop productivity. The world's present agricultural production system is based primarily on the needs of annual crops. There are a few exceptions—bananas, coconuts, other tree fruits and nuts, grapes, and a few vegetables, such as asparagus and rhubarb. Agricultural productivity varies dramatically from year to year. *Extremes in weather, rather than averages, affect agriculture. Crops and livestock are sensitive to weather over relatively short periods of time, and annual averages do not convey short-term differences.* Interannual variabilities have a far greater impact on agricultural productivity than does any current climate change. There is no evidence that with projected climate change, there will be increases in interannual variabilities.

Climate is the primary factor controlling the variability of crop yields. It also limits, in large part, the geographic areas for successful food crop production, the distribution of food animals, the growth of forests, species distribution and productivity of wildlife and fish. Annual variabilities in climate, whether extreme events (hail, drought, floods, frost, and freezes or climate-induced diseases) or seasonal variations in temperature, sunlight, or rainfall, result in significant year-to-year variations in food and fiber supplies. Of all the earth's planetary components the photosynthetically active biota are the most likely to be affected by the rising level of atmospheric CO_2. The first, which is already known, is the direct positive effect on the photosynthetic efficiency, and the second, on increased water use efficiency. These have been observed in thousands of experiments. The second would be a response to a climate change reflected in a global warming and, likely changes in precipitation patterns, which have not yet been confirmed.

Agriculture is directly responsive to these changes, and their global impact on food and fiber production will be reflected in future living standards (Schelling, 1983). Virtually all of agriculture everywhere is outdoors and directly exposed to the earth's atmosphere and climate. It is dependent on sun and water. Both crops and livestock are sensitive to temperatures, especially frosts, freezes, heat waves, and drought. They are subject to both the beneficial and harmful activities of insects, diseases, microorganisms, and weeds. All these are, in turn, affected by weather and climate. Rainfall, drought, hail, snow, and ice may destroy crops directly or result in leaching, salinity, and soil erosion. Food animals are dependent on weather and climate indirectly through the crops they feed on and the pests that afflict them, and directly through their sensitivity to temperature, sunlight, humidity, wind, rain, and snow. Each domesticated animal or species, as well as those in the wild have their thermoneutral zones (Ames, 1980; Hahn, 1976). Most farm labor functions outdoors in an unconditioned climate. Much the same can be said of forestry and fisheries.

Climates have been modified for crop production in diverse places. Deserts and semideserts have been made productive by irrigation. One third of the world's agricultural output and food supply is now grown on the 17% of the cropland that is irrigated. For the United States, irrigated land produces about one third of the value of crops on only a seventh of the harvested cropland (Waggoner, 1992). For specific crops, devices to create smoke and circulate heat may protect fruit orchards from frost; and overhead sprinkler irrigation systems may protect gardens, nurseries, orchards, and vineyards from frost and freezing temperatures; but they are too costly for most crops. Much of poultry production has moved indoors. Not only are temperatures modified and protection from the wind improvised, but the length of day is controlled. Hog production likewise is becoming industrialized, along with some dairy cattle herds. Greenhouses for high value crop production and plastic sheeting for soil mulching and plant covers (row tunnels) and greenhouses have greatly multiplied worldwide since the early 1970s. Large geographic areas comprising millions of hectares, heretofore climatically limited for crop production, have been expanded (Wittwer, 1993; Wittwer and Castilla, 1995). Weather modification in the way of precipitation enhancement may have had some limited success. With genetic engineering to increase the efficiency of photosynthesis, to make plants more resistant to frost and certain pests, and to facilitate the manufacture of protein and other products, food production may become less dependent on natural climate during the coming century. For the present, however, in most of the world, a land-intensive outdoor agriculture directly exposed to the earth's atmosphere and climate seems almost certain, as the dominant pattern for at least most of the 21st century.

Despite world agriculture's almost certain dependence on climate in the foreseeable future, it is not yet possible with today's crops, technologies, and the distribution of agriculture over the earth to assess or model the aggregate projected changing climatic impacts accurately. There is an almost universal, perhaps miscalculated assumption that any CO_2-induced climate change, independent of what the change may be, would be bad (Schelling, 1983). Even the positive direct effect of a rising level of atmospheric CO_2 on photosynthesis and water use efficiency has been interpreted as harmful when it comes to the growth of weeds, the feeding of some insect herbivores, interactions among forest tree species, arctic tundra, and most natural ecosystems. It is further espoused that all favorable responses to elevated levels of atmospheric CO_2 were made under controlled agricultural conditions in which all other environmental requirements were maintained at optimal levels. Specifically, it is stated that only when temperature, light, soil, water, and mineral fertilizers are present in optimal ranges, and when plants are not suffering from disease, weed competition, or herbivory, is there any significant response to elevated levels of atmospheric CO_2 (Bazzaz and Fajer, 1992; Houghton and Woodwell, 1989; Korner and Arnone, 1992; Strain, 1992; Strain and Cure, 1985). I do not accept these conclusions.

1.3 HISTORICAL ASPECTS

The issue of CO_2 has now been on the agricultural agenda for more than a century (Ausubel, 1983). This also applies to the first observations generated on the beneficial effects of elevated levels of atmospheric CO_2 on plant growth, development, and productivity. For over 100 years, nursery operators have been adding CO_2 to raise the yields of vegetables, flowers, and ornamentals grown in protective structures (Wittwer, 1985, 1986; Wittwer and Castilla, 1995; Wittwer and Robb, 1964). For decades, it has been known among biochemists, botanists, agriculturists, and foresters that a shortage of atmospheric CO_2 is often the most limiting factor preventing photosynthesis from proceeding more rapidly (Rabinowitch, 1956). It has only been within the past two decades, however, that these two possible impacts on crop productivity have been seriously brought together with some national and international consideration by climatologists and agriculturists. And only within the past decade has there been a serious attempt to relate the direct effects of rising levels of atmospheric CO_2 on plants with that of anticipated changes in the global climate (Adams et al., 1990; Lemon, 1983). As will be outlined, each, separately, or the two in combination, can have an enormous impact either for good or for ill on agricultural productivity and future food security. Unfortunately, Ausubel (1983), as well as most of his contemporaries, in his Historical Note (National Research Council; pp. 488–493, 1983) emphasized only negative impacts with no mention of the beneficial direct effects of rising levels of atmospheric CO_2 on plants. Meanwhile, he has been joined by reports from the World Resources Institute, the Worldwatch Institute (Flavin, 1989), and a flood of papers in *Scientific American* that either ignore or greatly minimize any mention of the direct effects of CO_2 on plant growth when making apocalyptic projections of so-called global warming.

Reference is also made by Ausubel (1983) to the seminal and oft-cited report of Revelle and Suess (1957), which emphasized only climate impacts with no beneficial direct effects. In a follow-up paper, however, Revelle (1982), often referred to as a technical optimist, emphasized that climatic impacts may be either beneficial or harmful, and increased productivity of such crops as rice, wheat, alfalfa, and soybeans should be enhanced by CO_2 enrichment. Meanwhile, Ausubel (1991a) has become a stalwart supporter of agriculture's ability to adapt to climate change.

The principal characteristic of climate is variability (Landsberg, 1984). Climate is the primary factor controlling the variability of crop yields in agriculture and its output. Variabilities in climate, whether extreme events of drought, floods, hail, frosts, or seasonal variations in temperature and precipitation, produce significant year-to-year variations in food supplies. *This makes stability of agricultural production as important, or more so, as the magnitude of production.* The uncertainties of climate make farm management alternatives and public policy programs for agriculture difficult to execute (Decker et al., 1985).

Climate shifts have occurred almost constantly during the earth's more recent geological history. Striking changes have occurred within the past 18,000 years. Then, the earth was at the height of the last ice age. Global temperatures were about 5 to 6°C lower than at present. About 15,000 to 8000 years ago, there was a rapid melting of the continental glaciers that covered Europe and northern North America. Antarctica and Greenland, however, remained ice covered. Since then conditions have been much warmer. About 5000 years ago, conditions became slightly cooler and in some areas drier. Civilization, as we know it, is a product of the past 10,000 years. Our technology and current projections of climate on food production and society are colored by adaptation to this relatively benign postglacial climate. Within the past millennium, mankind has experienced a warm medieval phase, a harshly colder episode between the 15th and 18th centuries (the "little ice age"), and a slow recovery to mildness in our own century (Hare, 1980; Newman, 1982, 1989).

Concerning the past millennium, two multiple century climate trends have occurred. A thermal maximum period occurred from about 900 to near 1300 A.D. During this period in the Northern Hemisphere, the Norsemen created permanent settlements near the southern tip of Greenland. It is estimated that annual temperatures rose 1 to 2°C, or even more, at these high latitudes. The Norsemen disappeared from Greenland in the early part of the 13th century. These warm centuries were followed by a thermal minimum period known in climatological literature as the "little ice age." It extended over nearly five centuries from about 1350 to 1850 A.D. The Northern tree line receded by about 300 km. Trunks of trees can still be seen in what is now Canadian tundra (Landsberg, 1984). The little ice age period had several minor rhythmic warming and cooling trends of a few decades each in length. Within the 20th century, temperature in the Northern Hemisphere rose slowly from the 1880s to a peak in 1938. Annual mean temperatures in the northern latitudes of 30 to 60° increased about 0.6°C during the 50-year period from 1890 to 1940 (Newman, 1982). Since 1940, a cooling trend approaching 0.5°C has occurred in many parts of the Northern Hemisphere at the higher latitudes. During the two decades preceding 1974, the United States experienced a period of exceptionally benign weather, resulting in a complacency that technology had advanced to such a level that weather or climate was no longer a significant factor in grain production (Thompson, 1975). The highly variable weather of 1974, and periodically in the years that followed has demonstrated that climate is still the most determinant factor in both the stability and magnitude of agricultural production. The current status of the global atmosphere is that the warming effect of the increasing atmospheric CO_2 could outweigh or overshadow the effects of the current cooling trend. There is also the possibility that the overall long-term cooling trend may be tempered by global warming. *History shows that warming is better than cooling.*

While climate change has been a natural feature of the earth's past, attention is now being given to predictions of climate effects of doubling the current atmospheric CO_2 level, since this could occur by the middle of the next century. For this situation, the various General Circulation Models predict an average global warming between 1.5 and 4.5°C and changes in global rainfall amounts and distribution (Bretherten et al., 1990). These possible climatic impacts and the direct positive effects of a rising level of atmospheric CO_2 on photosynthetic and water use efficiency, as world agriculture is impacted, constitute the primary focus of this book.

Climate has always varied. There is little or no evidence that overall variability has increased or that climate variability increases with climate change, even though there have been extreme anomalies in the

United States during the two decades preceding 1993 (Hare, 1980; Reifsnyder, 1989). These include the unusually wet spring of 1974 over most of the United States corn belt, followed by a summer drought in the Great Plains and then an early frost (Thompson, 1980), the drought and heat waves of the corn belt in 1980 and again in 1983 and 1988, and then a surprisingly cool wet summer in 1992. A 7-year drought in the western United States was followed by record breaking precipitation levels and floods in the west and central United States during the winter, spring, and summer of 1993.

Elsewhere, the Sahelian drought in the 1970s in Africa, for example, had several predecessors during the past 500 years. During 1972, there were droughts in Africa, India, the Soviet Union, and Australia, coupled with the failure of the Peruvian anchovy harvest due to ocean circulation changes. India was beset with a major drought in 1979, and China is perennially faced with droughts in the north and floods in the south. The United States corn belt was devastated in 1972 by what was called the southern corn leaf blight, a product of genetically vulnerable corn varieties and a series of weather anomalies. The yield of corn dropped by 50% or more in some southern states, and 15% nationwide (National Research Council, 1972; National Academy of Sciences, 1976) . Fluctuations from year to year have far exceeded any climate change that has occurred during the past century.

The difference today is that the human economy has become more vulnerable to climate variations. For centuries, pastoral nomads such as the Tuareg (Berber) and Fudani endured each Sahelian drought by migration to better grasslands with their flocks and herds, and in numbers of livestock. By endurance, Indian and Chinese peasants somehow survived countless failures of the monsoon. In North America, there was an adaptation to the ferocious winters of the late nineteenth century.

Today, human numbers have increased, especially in areas where food supplies are marginal. Old social and economic customs, notably nomadism, have largely disappeared. This is especially true in advanced Western societies, where there is the rising cost of energy, and a huge increase in the consumption of water. Demands for greater comfort, such as air-conditioning, are evidence of an increased sensitivity to climate anomalies (Hare, 1980). Conversely, it could be argued that for agricultural food production we have become less vulnerable to climate through the adaptations of irrigation, protected cultivation and housing for crops and livestock, shelter belts, genetic improvements for greater climatic resiliency, improved cultural practices, and the direct positive effects of rising levels of atmospheric carbon dioxide on the enhancement of photosynthesis and water use efficiency (Ausubel, 1991b).

2

Climate as a Resource in Food Production

2.1 INTRODUCTION

Climate is a resource to be used or a hazard to avoid. It is the first resource of all mankind. This major energy resource for the earth enters the human economy through photosynthesis: carbon comes from CO_2 in the atmosphere, water from precipitation via the soil, and energy from sunlight. From this remarkable feat of catalytic chemistry comes food, natural fibers, and timber. Climate is a resource even if we adopt the economists' "restricted" definition of the word. It is something scarce that needs to be allocated. It is a renewable natural resource. If climate is good for food production, recreation, health, or other purposes, it will command a high price. For example, in North America there are climatic sites in Michigan where the young fruit blossoms are never destroyed because of early spring freezes. This is true also of the temperature-moderating effects of the Great Lakes in western New York and Pennsylvania and in the Niagara peninsula of Ontario, Canada. Climate and its elements are also a set of resources we need to develop. Although it is renewable, it is not an inexhaustible resource. There are hazards that we either must cope with, adapt to, or learn to avoid. Whichever the approach, climate is a resource that has value, and that value will very likely change significantly if climate changes (Hare, 1980).

When we refer to climate, we mean the habitual state and behavior of the atmosphere, the gaseous envelope that covers the earth. The oceans and soils also have a climate as does the living cover of the earth, which we call the bioclimate. There is continuous motion and exchanges of energy from oceans and areas covered by snow and ice with the atmosphere. This creates a diversity of weather. Weather events over a period of time create climate. The climate is the sum or totality of all the weather experienced in the course of a year or more. We should think of it as long-range meteorology. In agricultural meteorology, we think of weather as being the real atmospheric conditions experienced in a particular place at a particular time (Elston, 1980). Climate is not only the conditions, such as temperature and precipitation, that can obviously be described as near average or normal, but also the extreme events (McQuigg, 1977). An arbitrary reference period may be 30 years or more. In fact, the World Meteorological Organization defines "normal climate" as the average of its measured site-specific daily weather elements or events, such as temperature and rainfall, for a 30-year period. The current normal is for the 30-year period from 1961 to 1990. There is no such thing, however, as a normal climate. History reveals that climate variations are a reality. It depends on where we put the reference point and how long we make it (Hare, 1980). Both climate variability and climate change have profound effects on the stability and magnitude of agricultural productivity (Rosenberg, 1988).

The real plague of agriculture is weather variability. In essence, this means rainfall and temperature variability. Temperature and water balance determine, in large, the distributions of crops, forests, and food animals. Production agriculture needs an improved predictability of weather within the time span of a crop's growing season, coupled with an improved predictability of season-to-season variability of weather and climate over a time span of a few years to a few decades. The real challenge is to determine the cause and effect of variability, rather than a response to one global perturbation, the so-called greenhouse effect (Allen et al., 1987). The need for and implementation of an agricultural weather information service is hereby called for and will be emphasized later.

The following covers the essential elements of climate that impact on food-producing systems.

2.2 TEMPERATURE

Temperature and precipitation, separately and in combination, with variabilities and extreme events associated with them, are the most common elements used to describe the world's climate They depict the major trends and variabilities of past climates. They also indicate what future climate and weather events might be.

Air temperature is probably the most common climate determinant of agriculture. It is also the most widely used atmospheric indicator of short- and long-term climate fluctuations. Temperatures largely control the rate of plant growth, flowering and fruiting responses, seed development, the water vapor flux, plant water status, soil drying, and irrigation practices.

Plants, unlike warm-blooded animals, have no mechanism for automatic regulation of oxidation and transpiration processes, making them incapable of maintaining a constant temperature. Whereas each animal species has its "comfort zone", all life processes, including those in plants, are restricted to a certain "biokinetic" range of temperatures. Both above and below this range, all organisms suffer a more or less rapid, and more or less reversible chill or heat injury. For photosynthesis, the range for reversible change extends for plants adapted to moderate climates from about 0 to 40°C. Because of evaporative cooling leaf temperatures of plants grown in irrigated arid climates may be 5 to 10°C below air temperatures. For land plants in moderate climates, the optimum is 30 to 35°C. For plants adapted to cool temperate and cold regions (Parry et al., 1988a), it is much lower. It is much higher for tropical semiarid desert plants (Parry et al., 1988b). Below 5°C and above about 25°C, a slow chill injury or a slow heat injury may set in for some plants. Then the rate of photosynthesis depends not only on the prevailing temperature but on the length of the exposure (Rabinowitch, 1956). The changes may also be irreversible. Temperature anomalies occur when heat waves kill chickens and blizzards starve flocks and herds. Crops are also sensitive to weather and climate. This is reflected in frosts that kill vegetables and citrus, and destroy the fruiting buds of deciduous fruit trees, and high temperatures that destroy rice and wheat fields and prevent pod-set in beans.

The rising levels of atmospheric CO_2 are commonly and almost universally predicted to increase air temperatures. It is imperative to assess the direct effects of increasing atmospheric CO_2 which will interact with concomitant warming.

There are many components of temperature that could impact on global agriculture and future food security. A brief description of some of these elements follow:

2.3 LENGTH OF THE GROWING SEASON

For temperate zone agriculture, and in most all of the northern and southern hemispheres beyond the tropics, it may be expected, but not guaranteed, that a warmer climate will lead to an increase in the length of the frost-free period, irrespective of altitude or latitude. This should also hold for agricultural enterprises in the mountainous areas of the tropics. The correlations between temperature and the length of the growing season also mean that with warming there will be a gradual shift northward of the northern boundary for such crops as winter wheat and corn in North America, Europe, and Asia. A comparable shift would be southward in the Southern Hemisphere. Similarly, the risk of freezing temperatures in Florida, Texas, and California in the United States may be reduced under the climatic regimes of a global warming. The same would also hold for southern Europe, all Mediterranean countries, subtropical parts of Asia, Africa, and the countries of the southern hemisphere. Crops could be planted earlier and harvest periods extended. This is assuming that a CO_2 induced warming would not increase climate variability and the risks to agricultural production. On these issues vital to agriculture productivity, climate modelers are noticeably silent, and modeling has not given answers, although some pronouncements have come forth with little supporting evidence. That variability that has occurred in the U.S. corn belt during the past two decades, following two decades of essentially benign climate during the 1950s and 1960s is no model for the future (Landsberg, 1984; Thompson, 1975, 1980). As little as a one degree increase in average temperature would not only lengthen the growing period by a week to 10 days, but if sustained over several seasons, could allow farmers who now grow just two crops a year to grow three, or where one grew before, to now grow two. Further details concerning the lengths of growing seasons on agricultural and forest productivity will be discussed in Chapter 3, Section 3.5.

2.4 INTERANNUAL VARIATIONS

Historically, climate variability has had a greater impact on agricultural productivity than has climate change. History tells us that the magnitudes of interannual variations exceed those of long-term trends. This pattern will likely continue. There is no evidence that variability will increase with the effects of rising atmospheric gases on climate changes (Reifsnyder, 1989). Moreover, the detection may be easily masked by year-to-year variability.

The summers of 1988 and 1992 provide an excellent example of the impact of a very recent 4-year apart interannual variation on agricultural productivity in the corn belt and bread basket of the United States (Table 2.1). The summer of 1988 was hot and dry and 1992, cold and wet. Mean monthly temperatures in some states and for some months, between the two years, differed as much as 10°F, and the contrasts in

TABLE 2.1. Temperature and Precepitation Differences in Corn Belt States of The United States in 1988 and 1992

State	IL	IN	IA	KS	KY	MI	MN	MO	NE	ND	OH	SD	WI
Temperatures of Corn Belt States—1988 Average 74.3													
June	73.4	72.0	74.1	77.2	72.7	66.3	71.1	75.2	76.6	74.1	69.1	74.7	68.9
July	77.5	76.4	75.8	78.0	77.7	71.9	73.3	77.2	74.6	72.2	75.7	74.7	72.9
Aug	78.0	78.0	76.2	79.1	77.8	70.5	70.3	78.6	74.5	69.1	74.4	74.4	71.7
Av.	76.3	75.4	75.4	78.1	76.1	69.5	71.6	77.5	75.2	71.8	73.1	74.6	72.0
Temperatures of Corn Belt States—1992 Average 67.0													
June	68.8	66.4	67.7	68.6	69.1	60.7	62.4	70.3	65.8	61.5	65.0	63.2	62.1
July	73.4	72.4	68.7	67.0	76.1	63.6	62.7	76.2	68.9	61.8	72.0	63.8	64.1
Aug	73.4	67.8	66.1	70.7	71.0	62.9	63.3	70.3	67.2	62.9	67.2	64.4	63.6
Av.	72.0	68.9	67.5	68.8	72.1	62.4	62.8	72.3	67.2	62.1	68.0	63.8	63.2
Precipitation in Corn Belt States—1988 Total 129.15 Average 9.93													
May	1.79	1.53	1.54	2.72	2.87	1.00	2.16	2.33	3.97	1.85	1.73	3.46	1.14
June	1.05	0.74	1.71	1.92	0.96	0.91	1.62	1.39	2.27	1.80	0.86	2.40	1.44
July	2.65	4.26	2.24	3.29	5.05	2.92	2.02	4.58	3.28	1.45	4.35	1.29	2.49
Aug	2.49	3.13	4.12	1.84	3.77	4.18	5.42	3.60	2.69	1.40	3.44	2.09	4.04
Total	7.98	9.66	9.61	9.77	12.65	9.01	11.22	11.90	12.21	6.50	10.38	9.15	9.11
Precipitation in Corn Belt States—1992 Total 183.24 Average 14.10													
May	1.19	2.38	1.73	2.24	4.05	1.12	2.17	2.39	1.69	1.70	2.84	1.47	1.87
June	1.66	2.80	1.84	5.58	4.21	2.07	3.75	2.95	3.49	3.03	2.63	5.54	1.94
July	7.25	8.15	8.35	7.45	6.04	4.65	4.06	6.80	5.30	2.57	8.12	4.59	4.05
Aug	1.84	2.84	2.39	3.71	3.67	2.79	3.98	2.30	3.36	1.95	3.51	2.43	2.76
Total	11.94	16.17	14.31	18.98	18.42	10.63	13.96	14.44	13.84	8.80	17.10	14.03	10.62

precipitation during the growing season varied as much as 4 to 7 in. Overall, corn yields for the United States achieved a record 132 bushels/acre in 1992, compared to only 85 bushels in 1988. For North Dakota, the same area was planted in 1987 and 1988, but in 1988 only 78% of the area was harvested and production went down to only 38% of the previous year. In 1992, corn yields for the United States exceeded by over 55% those in 1988. Again in Iowa, the same area was planted and harvested in 1987 and 1988, but only 69% as much corn was produced in 1988. In 1992, an all time record high of 9.48 billion bushels of corn were harvested in the United States compared to 4.92 billion bushels in 1988. The overall climate impact on the reduction of corn yields and total production in 1988 exceeded that of losses from the highly publicized southern corn leaf blight of 1972 (National Research Council, 1972). Nationwide, July 1992 was the coolest since 1950 and the wettest since 1958. Temperatures averaged 6°F below the so-called 30-year normal.

Even more recent historical examples of interannual variations in the United States was the flooding and water-logging of soils throughout the central corn belt in 1993, and in Georgia with tropical storm depression Albert in 1994.

Interannual variations in temperature and precipitation have an astounding impact on the agricultural production stability as well as on the magnitude of crop production. Thompson (1975) emphasized that most of the world grain production is in the middle latitudes, where summer temperatures average between 70 and 75°F (21 and 24°C). The limitations are high summer temperatures at the lower latitudes, and length of growing season at higher latitudes. Thompson (1975) further emphasized that the highest yields of corn and soybeans seem to occur in years of lower than normal temperatures in July and August and higher than normal rainfall. This was admirably confirmed once again by the contrasting production in 1988 and in 1992.

2.5 EXTREME EVENTS

There are climate factors other than changes in temperature and precipitation that impact interannual variabilities and agricultural productivity. These are the so-called "extreme events". Included are high winds, dust storms, ice storms, hail, hurricanes, typhoons, tornadoes, and floods, which each year devastate large food-producing areas, especially in the tropics and subtropical regions of the northern hemisphere. Hurricanes Gilbert (1988) and Hugo (1989) and the floods of the Midwest (1993) are recent reminders in the United States. Two of these extreme events for North America occurred during one of the hottest, driest (1988) and one of the coldest, wettest (1992) years in the United States during the past several decades. Air stagnation is becoming increasingly important where food-producing areas border on, or are within industrialized areas.

Volcanoes can have a marked cooling effect on the entire globe. In fact, cloud cover is many times more powerful in affecting temperatures than the greenhouse gases,and is infinitely more variable (Brooks, 1989). The most recent global cooling was from the eruption of Pinatubo in the Philippines during the summer of 1991. It is suggested that the sun-blocking aerosol clouds resulted in a 0.5°C temperature drop for the entire earth during 1992. It was the most massive eruption since that of Krakatau in 1883 (Kerr, 1993a). Volcanic activity, like other climate events, is unpredictable. Killing frosts, freezes, floods, and droughts occur almost every year, and sometimes several times a year, in one or more locations in the United States. The same is true of other major food-producing areas in Europe, Western Russia, India, China, Argentina, and Australia. The land areas in the United States that have environmental limitations include those with low water availability, comprising 45% of the land surface, and those too wet (16.5%) or too cold (15.7%). Similar data apply to the soil surfaces of the world. The effects of unfavorable climates are as pervasive as those of unfavorable soils. This relates to the total indemnifications paid to farmers for crop losses. Drought, excessive water, and cold account for 71% (Boyer, 1982).

The notorious urban heat island effect, most extensively measured for the United States, has raised urban temperatures by 1 to 2°C or more, and may exceed the rural surroundings on occasion by 10°C (Landsberg, 1984). Whereas this may not be an "extreme climate event", the rates of temperature increases are equal to, or exceed, the extreme of those projected for global warming being induced by doubling of the current levels of atmospheric CO_2 and other gases. In fact, Changnon (1992) suggested that the ongoing urban heat islands and adaptations relating thereto are a useful historical and partial analog for assessing the impacts of future global climate change on food production, human health, and economic and social outlays. The impacts on agricultural productivity may not be as immense as those on human health and the energy budget, but much of the world's agriculture and some important food-producing systems are within the boundaries or adjacent to growing urban heat islands. The impact of heat islands and impacts on temperature records will be further discussed in Chapter 3.

2.6 GROWING DEGREE DAYS

Thermal units, commonly referred to as heat units in agricultural production, can be an intricate measure of temperature-sensitive biological phenomena relating to crop maturity, dry matter increases, pest migrations, and plant disease infestations.

The rates of maturity of most major food crops, are highly temperature dependent and sensitive. For example, times of maturity for both peas and sweet corn, extensively grown for processing in the United States, have large data bases that give heat or temperature requirements for maturity as determined by tenderometer readings. Data for the past 50 to 60 years are available from food-processing companies for specific varieties and locations. Thermal units expressed as growing-degree-days are a very sensitive measure of climate change. They are available for entire growing seasons. A continuous variable, such as temperature, is likely to be more sensitive than the discontinuous variables of precipitation and spring and fall frosts. The sensitivity of the temperature response is that maturity for a crop for a given variety and location can be accurately predicted as a function of accumulated temperature called growing-degree-days. The baseline is above 4°C for peas and above 10°C for sweet corn.

Assuming there is a modest global warming in temperate zone agriculture, heat sums above 10°C at higher latitudes will increase for a given growing season. Budyko (1982) suggested that this may be as much as 1500°C in what was the Soviet Union. If this were to occur, many of the areas for the production of the major food crops, such as corn, wheat, rice, and potatoes, could be greatly expanded northward. The base temperature for most of the major food crops is at, or slightly above, 10°C.

2.7 WATER RESOURCES

Vital to future food supplies and total agricultural output is water. Of all the earth's natural resources, and as a component of climate reflected in precipitation, it will become the most critical. Any degree of global warming will increase the demands for water in crop production and, to a lesser extent, for livestock and other food animals. The prospect of climate change heightens the likelihood of water and regulations (Waggoner, 1990). Water will have to come either from increases in precipitation or in the existent surface or underground reservoirs, from greater water use efficiency by crops, or from improved irrigation management.

By far the most extensive means of protected cultivation of crops on earth is through irrigation. By this means crop production has been extended to deserts and semi-arid lands. Many of the hazards of drought have been overcome. The rate of increase for irrigated land has been precipitous. From 1950 to the early 1980s, the crop area irrigated in the world increased from 94 to 250 million ha (Postel, 1989). This increase in irrigated land accounted for 40 to 50% of the increase in agricultural output. Although the expansion of irrigation in the late 1980s has slowed dramatically in the United States and elsewhere, some nations such as China, anticipate further expansion. *Today, the 17% of the world's cropland that is irrigated produces one-third of the agricultural output.* Over two thirds of all the food in China is produced on cropland that is irrigated and more than half of the food produced in India, Indonesia, and Japan is grown in irrigated fields. Without irrigation, there would be essentially no food produced in Egypt, the Sudan, Israel, and Saudia Arabia. For the world's five most populated countries—China, India, Russia, the United States, and Indonesia, water supplies for agriculture are increasingly becoming more critical. In the United States, the 12% of the cultivated farm land that is irrigated now accounts for 37% of crop production. Agriculture in the United States consumes, mostly through irrigation, 80 to 85% of fresh water resources. Worldwide, agriculture claims two thirds of all the water removed from rivers, lakes, streams, and aquifers (Postel, 1993). This is true for most countries of the world, except those in northern Europe where drought seldom occurs and irrigation is infrequent. Further agricultural expansion is now limited by short water supplies in much of the United States, Russia, China, India, Pakistan, Spain, Italy, Turkey, Mexico, Chile, Australia, and almost all Mediterranean, Arabian Gulf, and African countries. The high agricultural use of water is increasingly under challenge, both from the standpoint of the rising costs of food and from other competitive needs, such as recreation, municipalities, industry, energy generators, and for the current low efficiency (35 to 40%), (Postel, 1993) of agriculture use. Making crop irrigation more efficient must be a top future priority. One option is the increased use of micro-irrigation techniques (Smajstria, 1993), which give up to 95% efficiency. Only about 1.5 million ha worldwide use water this efficiently.

Water is both our most precious and most wasted resource. As a resource, it is being transferred from farming to other uses. Water scarcity is a spreading global problem. Whether climate changes or not, this poses a challenge for future food production. If there is global warming, ease of adaptation to a warmer and drier climate will likely be retarded by a lack of water. An atmospheric CO_2 induced climate change, coupled with possible impacts on global water resources relating to crop needs, and an enhanced water use efficiency for world food production, poses a remarkably complex set of interacting factors not yet represented by any predictive climate or plant growth models.

2.8 PRECIPITATION PATTERNS

It is not only the annual amount of rainfall or snow cover that occurs at a given location for specific crops that is pertinent, but its distribution. In the United States, the corn, soybean, and wheat belt and most other important grain-producing areas in the middle latitude have summer temperatures which average between 70 and 75°F (21 and 24°C). Temperatures for rice and soybean may be slightly higher than those for wheat. Marked deviations above this range, especially if accompanied by lack of rainfall, will greatly reduce yields. This occurred in the United States in 1988. Corn and soybean yields in the United States are associated almost in a linear fashion with the July and August rainfalls. Yields of wheat grown in the Great Plains of the United States are best when rainfall is greater than normal. In Indiana and Illinois, yields are greater when rainfall is normal or slightly below normal in late spring and early summer. Higher than normal temperature, from flowering until the crop is mature, adversely affects wheat yields throughout the United States. Adequate moisture is critical for all cereal grains, legumes, and fruit crops during flowering, seed formation, and fruit setting. Supplemental irrigation during these periods of crop development in otherwise humid or semihumid areas is often of great value for yield.

2.9 IRRIGATION

The merits of irrigation lie not only in improving the stability of crop production, but also in its magnitude. It also accounts for the two to three times greater output of crops grown per unit land area, under irrigation compared to the rainfed or dryland.

Irregularities in precipitation not only have an immediate effect on crops, but may have a more lasting impact on reservoirs and storage basins used for irrigation. Massive diversions of water for crop irrigation, especially if coupled with a decrease in precipitation, may not only irreversibly reduce surface supplies of fresh water but have many other environmental impacts. An example is in Russia (now Kazakhstan) with the depletion of water of the Aral Sea.

Egypt and The Sudan have about half the irrigated land in all of Africa. The construction of the Aswan Dam is dramatically changing not only the availability of water for irrigation but the nature of the water. The Nile river is now essentially a closed system with no outlet into the Mediterranean. Boulder Dam in the southwest United States has greatly improved the availability of water for irrigation and other purposes, but has greatly modified the ecology downstream. There is concern, however, that future water supplies will be adversely affected by prolonged drought and the progressive lowering of Lake Mead. Expectations for water appear to exceed the river's ability to meet them. The long-term estimated one billion dollar Mahaweli Ganga and Walave

Ganga irrigation project in Sri Lanka has completely altered the agricultural potential and energy resources of that tropical country. Thirty-nine percent of the total land area will be affected by this project. China is anticipating a massive water diversion and reclamation project for the Yangxi River, which would dislocate millions of people and submerge the Three Gorges, but bring millions of new hectares under irrigation and be a major source of hydroelectric power. This will circumvent the irregularities in precipitation for much of the North China Plain and greatly improve the stability and dependability of food production for a nation which has 22% of the world's people, but only 7% of the arable land (Wittwer et al., 1987).

2.10 DROUGHT

The lack of water is the single most important impediment to plant growth and global food production. The greatest crop losses in the United States from 1939 to 1978 were caused by drought. Losses from drought were about equal to all other climate-induced losses including excess water, cold, hail, and wind (Boyer, 1982).

Drought is feared by farmers throughout the world more than any other climate extreme. It is a periodically recurring phenomenon throughout the world (Idso, 1989a), and the United States is no exception. Major droughts during the 1980s occurred in 1980, 1983, and 1988. These droughts were accompanied by major heat waves and resulted in substantial crop losses and very significant decreases in yields of both corn and soybeans. It is very significant to note, however, that the percentage decreases of yields of corn, wheat, and soybeans in the U.S. grain belt during the droughts of the years 1980, 1983, and 1988 were progressively less than those experienced in the 1930s, 1950s and 1970s. This was not a result of a decrease in severities, but of increased inputs from new technologies. There were recurrent droughts in the south central United States from 1750 to 1980. There are those still living who remember the dust storms in the Great Plains in the 1930s. This was a decade of dryness, with 2 years (1934 and 1936) of extreme drought (Rosenberg, 1980). Corn yields that had previously averaged near 25 bushels/acre fell to 10, 5, or even 2 or less bushels per acre in some areas. But droughts did not begin in the 1930s. The skies of the eastern United States from what is now the midwest corn belt were darkened by dust storms and drought in 1716, 1762, 1780, 1785, and 1814 (Schultz, 1982). During the postresettlement period, prior to the 1980s, there were major droughts in the 1860s, the late 1880s to mid-1890s, the 1930s, the mid-1950s and the mid-1970s (Rosenberg, 1980).

Droughts, localized or widespread, are a common occurrence in all of the larger food-producing countries and major geographic areas of the earth such as the United States, India, China, the Middle East, Australia, and Russia. While northern Europe is relatively free from the climatic extremes of hot dry weather, there was a major drought and heat wave in the early 1970s and again in 1994. In Russia, there are millions of hectares of land marginally cold and dry. Hence, the Russians cannot consistently feed themselves. Major droughts have plagued the Middle East and go back to Biblical times, highlighted in Egypt by the 7 years of plenty followed by the 7 years of drought. The pervasive drought in the Sahel of Africa during the 1970s is still in our memories. There were droughts in India during the mid-1960s when the monsoons failed, and again in 1979. China, in its over 40 centuries of farming, has periodically suffered from famines induced by droughts, especially in the northern and western provinces. The most recent drought was in 1992 when the spring and summer months were extremely dry in five of the southeastern provinces that normally receive adequate rainfall. Again, the interannual variations for temperature and precipitation equal or exceed the greatest extremes projected for a doubling of current atmospheric CO_2 levels.

The prevalence of drought has prompted interest in increasing water resources through the augmentation of precipitation through cloud seeding. There is, however, no area of technology so vulnerable to regulatory constraints. Any successful technologies for rainfall enhancement will have numerous socioeconomic political, legal, and environmental impacts. Although some precipitation enhancement research using new cloud-seeding technologies, both in the United States and the former Russia, suggests very positive results for agriculture, the value has not yet been scientifically determined. In a comprehensive review in the American Association for the Advancement of Science document, "Climate Change and United States Water Resources" (Waggoner, 1990), weather modification through cloud seeding for increasing precipitation is not given as an option. This is irrespective of the conclusion that water resources will become the most critical limiting resource in any global warming episode.

2.11 THE CHEMICAL CLIMATE

Chemical climate means certain gases (chemicals) in the atmosphere absorb some of the outgoing infrared radiation from the earth's surface. This absorbed terrestrial radiation, which is being declared as a function of the concentration of certain greenhouse gases, is causing global warming. This absorptive capacity for radiation, the so-called "greenhouse effect", is thus related to the chemical composition of the atmosphere and has an impact on climate, particularly the temperature of the atmosphere. The intriguing feature of the chemical climate is that the chemical composition of the atmosphere is believed to be changing largely as a result of anthropogenic inputs. The most important of these chemical inputs is CO_2, which is increasing at the rate of about 0.35% per year. The total increase since the beginning of the industrial era is about 30%. It is by far the most important ingredient in the chemical climate and is estimated to account for over 50% of the radiative forcing.

Methane (CH_4) is a gas that contributes to the chemical climate. It has been increasing at the rate of about 1% per year for the last decade or so and at a rate of about three times that of CO_2. A significant downward departure from this rate of increase has, however, recently been noted (Kerr, 1994a). Methane contributions to radiative forcing during the decade of the 1980s are estimated at 15% (IPCC, 1990).

Nitrous oxide (N_2O) is also a part of the chemical climate that is changing. It has increased from its preindustrial atmospheric level of about 288 ppb to 310 ppb in 1990. The current rate of increase is 0.25% per year. Although it is accumulating at a rate more slowly than CO_2 and much more slowly than CH_4, its effective warming potential per unit emitted will be, after 20 years, much greater, perhaps 300 times that of CO_2 (IPCC, 1990). It is now contributing about 6% of the radiative forcing of the atmosphere.

The chlorofluorocarbons (CFC-11, CFC-12) are also found in the atmosphere at 280 and 484 ppb (1990 values, respectively), and they are increasing at the rate of approximately 4% per year. They are considered major contributors to lower levels of observed ozone in the stratosphere (IPCC, 1990), are strongly absorptive of longwave terrestrial infrared radiation, and are estimated to contribute about 17% of the radiative forcing of the atmosphere. A complete accounting of all gases contributing to the chemical climate of the atmosphere, with a master table of them and their attributes, is detailed by Wuebbles and Edmonds (1991).

It is important to emphasize that the evolution of trace gases, in terms of radiative balance or atmospheric chemistry, has been documented only for very recent periods. For CO_2, this has been since 1958, and for methane only since 1978. There has been no increase in these trace gases in the Northern Hemisphere since 1990 (Kerr, 1994b). The three greenhouse gases, CO_2, CH_4, and N_2O, are now increasing primarily as a result of anthropogenic influences. The current mean atmospheric concentrations approach 360 ppmv (parts per million by volume) for CO_2, 1700 ppbv (parts per billion by volume) for CH_4, and 310 ppbv (parts per billion per volume) for N_2O. The levels just before the time of the major anthropogenic inputs as revealed by ice core records were 280 ppmv for CO_2, about 700 ppbv for CH_4, and 260 to 280 ppbv for N_2O (Raynaud et al., 1993). Although all these gases contribute to global warming, it must be remembered the most abundant greenhouse gas in the atmosphere is water vapor. It too will go up if warming occurs.

2.12 SUNLIGHT, SOLAR ENERGY, AND CLOUD COVER

Chemically, green plants are the only productive part of the earth's population of living things. They are self-supporting ("autotrophic"), and they alone enable people, animals, fungi, and most bacteria to exist. Green plants accumulate chemical energy. They add to the food and chemical resources of the earth. Other organisms dissipate them. The reduction of CO_2 by green plants is the largest single biochemical process on earth. The yield has been estimated at 10^{11} tons/year, which daily produces over 100 lb of cellulose for every one of the earth's 5.7 billion people. Cellulose is the earth's most abundant plant product.

The key to this most important of biochemical processes is solar energy. Green plants are biological sun traps. They capture energy from the sun and produce a multitude of useful products, one of the more important being food. Light saturation for photosynthesis of individual leaves occurs at an intensity close to that of full sunlight, at high noon, at middle latitudes. This is equivalent to about 10,000 foot candles (Rabinowitch, 1956). Direct comparisons, however, of foot candles of sunlight with other sources of light are not valid without a knowledge of their spectral energy distributions. Some food plants, such as the C_4 plants, to be described in detail later, include corn, sorghum, millet, and sugarcane. These crops progressively respond to sunlight intensities up to the maximum available. Others such as the C_3 plants, like wheat, rice, soybeans, potatoes, cotton, and forest trees having whole plant canopies, will usually plateau at levels below that of full sunlight. Sunlight, along with precipitation and temperature, is a unique component of climate. Its intensity and duration will very likely be altered in any proposed scenarios. Yet, most all scenarios take no account of a possible major change in the energy output of the sun, such as cloud cover, fog, smog, and volcanoes, which in turn can have a marked effect on global temperatures and exert both direct and indirect effects on our food supply. *Clouds affect photosynthesis as well as temperature.*

Contrary to the beliefs of many ecologists and climatologists, ideal conditions for CO_2 fixation for agricultural crops, during the growing season or part thereof, seldom exist. Most agricultural areas are marginally too cold or too dry or too hot or too wet. Sunlight is almost always limiting, except in desert areas, for optimal photosynthesis to occur. This is particularly true of greenhouse crops grown during the fall, winter, and spring in temperate zones. Light is a limiting factor for rice grown in the lowland humid tropics, where frequent cloud cover, especially during the monsoon rains, greatly restricts growth. It is true of corn fields in Iowa, trellised crops in China, corn, soybeans, and rice grown in Indonesia, trees in a temperate zone forest canopy, and in the rain forests of the tropics. Ideal growing conditions,such as equal parts of water and sunlight seldom exist, except in the few instances where the rains come at night, and there is full sunlight during the day. In addition, there are throughout the world many soils subject to environmental limitations, such as those that are cold and wet, alkaline, saline or shallow, or have nutrient excesses or limitations (Boyer, 1982). Seldom, indeed, are temperatures, soil nutrients, and sunlight taken alone, or in combination either in the field or a greenhouse, optimal for plant growth.

Photoperiod

Photoperiod, or the length of daylight, while relating to location perhaps more than climate, has a significant impact on the responses of many, if not all, crops. Day length interacts with temperature, precipitation, and sunlight. Most all crops, including the major food crops—rice, wheat, maize, potato, sweet potato, peas, and beans—and tree fruits, as well as forest trees and shrubs, respond to the length of days not only in vegetative growth, but their reproductive development in producing flowers, seeds, bulbs, roots, and tubers. Rice, the number one food crop of the earth, during the stress periods of panicle differentiation, flowering, and grain filling, is sensitive not only to temperature and moisture but to day length as well. If this is accompanied by a global warming, and if cropping areas and habitats migrate or move in latitude, photoperiod as a variable, relating to crop productivity, must be dealt with. Many native species and varieties of crops, such as the soybean, have very specific photoperiodic requirements. Similarly, specifications of temperature dependence, interacting with photoperiodic limitations for vegetative extension, flowering, fruit setting, bulb formation, and breaking of dormancy of many horticultural crops, root, bulb, and tuber crops, cereal grains, and forest trees may define the limits of habitat both in latitude and altitude for many crop species. If there is a global warming, successful deciduous fruit tree and winter wheat production would move northward. One of the results would be a change in photoperiod.

Thermoperiodism

An important response of plants to a temperature variable, described over 50 years ago (Went, 1944, 1957), is that of thermoperiodicity. Plant growth and development are sensitive to both day and night temperatures independently, as well as the differential between them (Mooney et al., 1994). The literature that is now available reports that seed germination, flower initiation, fruit setting, bulb formation, and tuberization may be particularly sensitive to day/night temperature symmetry. Some of the best examples are with horticultural crops, as well as for rice and soybeans. It was demonstrated, for example, that night temperatures controlled the flowering and fruit setting of tomatoes (Went, 1944). The effective range was between 15 and 20°C. Night temperatures either below or above resulted in a poor crop (Wittwer et al., 1948). Height in bedding plants is affected more by the differential between day and night temperatures than the absolute temperatures (Erwin et al., 1989). Cold exposure during early daylight hours was found to be an effective means of controlling plant heights. For tropical rice, seed yields are linearly decreased 10% per °C at temperatures from 26 to 36°C irrespective of atmospheric CO_2 levels (Allen et al., 1989a; Baker and Allen, 1993a, b). Soybeans produced no seeds at day/night temperatures of 18/12°C at any CO_2 level (Rose, 1989). Plants apparently utilize morphogenetic events (Mooney et al., 1994).

Volcanoes

The climatic effects of volcanic eruptions are well documented (Bradley, 1988). The global environment and food production, both regionally and globally, may be dramatically impacted by volcanoes. Two examples are given—one from ancient records and the other in modern times. Correlations of archeological records and soil-stratigraphic data with geological events and socioeconomic responses reveal that there was an abrupt climate change in approximately 2200 BC. It resulted in the collapse of the third millennium North Mesopotamian civilization of the Akkadian Empire. There was a marked and sudden increase in aridity and wind circulation following a volcanic eruption. It stayed dry for about 300 years (Weiss et al., 1993).

During June 1991, Mount Pinatubo in the Philippines sent more volcanic aerosol into the earth's stratosphere than any tropical eruption since Krakatau in 1883, and more than twice that of any other recorded blast in the past century. Liscio (1993) correlated the Pinatubo eruption in 1991, with a global temperature in 1992 being colder than any time since the early 1920s, the United States experienced the coolest summer in 77 years (Kerr, 1993a). The entire United States was the wettest in 1993 that it had been since 1983. Other possible effects of the Pinatuba eruption have been a marked reduction (18%) of atmospheric carbon monoxide between June 1991 and June 1993, and a dramatic slowing of the rising atmospheric CO_2 beginning in July 1991, which did not return to its normal pace until mid-1993 (Kerr, 1994b). Still other correlations are the U.S. floods of 1993, the most devastating in recorded history; Nepal experienced the worst deluge in 60 years, leaving thousands dead; and millions homeless in northern India and Bangladesh. Added to the above was Hurricane Andrew, recorded as "the nation's greatest natural disaster in its history," devastating South Florida in September 1992, and the heavy snow storm of March 22, 1993, which the U.S. National Weather Service called "the single biggest storm of the century," sweeping from Florida to Maine (Petranek, 1993). *Volcanoes can devastate the climate.*

2.13 PROTECTED CULTIVATION—CONTROLLED ENVIRONMENT AGRICULTURE

Among the greatest constraints in agricultural crop production are a lack of sunlight, adverse temperatures, droughts, weed growth, and deficiencies in soil nutrients and atmospheric CO_2. Most of these are climate factors or directly related thereto.

Food production and agriculture in general are not as vulnerable to climate and climate variations today as in generations past. The single most important technological input for controlled environment agriculture has been the phenomenal global expansion of irrigation. Crop irrigation has risen from 94 million ha in 1950 to over 250 million ha by the early 1980s, resulting in a third of all agricultural production coming from irrigated cropland. About 17% of the world crop acreage is now under irrigation. Most rice is produced on irrigated, or other water-managed land. Many desert areas have been reclaimed. Two or more crops can be grown per year where only one was grown before.

The benefits from irrigation are twofold: greater stability and dependability of production and significant enhancements of the magnitude of production.

Controlled environment agriculture has now extended far beyond the realms of crop irrigation and water management (Wittwer, 1993; Wittwer and Castilla, 1995). The Chinese have long employed both natural and man-made structures as windbreaks for crops. In recent years, this has been furthered by windrows of trees circumscribing millions of hectares in the northern provinces to protect against the arctic gales in winter and the hot searing winds of summer. Windbreaks effectively reduce evaporation and evapotranspiration, modify daily temperature variations, and reduce sand motions in areas bordering deserts.

Plastic films or mulches are laid over millions of hectares of soil in China and other northern Oriental countries, where crops of cotton, peanuts, rice, corn, watermelon, and many vegetables are now grown. Globally, in 1992, there were 3 million ha of row crops (cotton, peanuts, corn, tobacco, rice seed beds, and many vegetables) grown in soils covered with plastic films. White plastic reduces soil temperature. Black increases soil temperature. These climate modifications result in extensions of cropping areas into new farming frontiers. Soil moisture is preserved, soil temperatures are increased, thus prolonging the growing season sometimes by 4 to 6 weeks, and weeds are controlled. Plastic soil films or mulches are often combined with plastic or other covers to form row covers. They may be expanded in size to form greenhouses, which in China alone, now cover some 75,000 ha, and worldwide over 200,000 ha. Such structures, some heated and some not, greatly extend the growing season and enable more crops to be produced each year. Rice seed beds covered with plastic film during spring in many Oriental countries extend the growing season by a full 30 days, increase the harvest index, and bring into grain production hundreds of thousands of hectares of new lands (Table 2.2).

What has happened in China during the past decade in the expansion of controlled environment agriculture was preceded by an enormous expansion of protected cultivation in Japan and Korea, in northern Europe, Israel, and, to a lesser extent, in the United States and Russia. Parallel with, and slightly preceding the expansion of protected cultivation in China, have been remarkable expansions in the Mediterranean countries of Spain, Italy, Greece, Turkey, Morocco, Jordan, and Egypt. Here man-made, as well as natural windbreaks are combined with plastic greenhouse covers for the production of vegetables, such as tomatoes, cucumbers, peppers, eggplants, green beans, and melons

Table 2.2. Global Distribution of Protected Cultivation (Modified after Wittwer and Castilla, 1995)

	Geographical Areas				
Structures	**Orient**	**Mediterranean**	**North and South America**	**North Europe**	**Total**
Plastic soil mulches	3,080,000[a]	191,000	85,000[c]	15,000	3,371,000
Direct cover (floating types)	5,500	10,300	1,500[c]	27,000	44,300
Low tunnels (row covers)	143,400	90,500	20,000[c]	3,300	257,200
High tunnels		27,600[b]			27,600
Plastic greenhouses	138,200	67,700	15,600	16,700	238,200
Glass greenhouses	3,000	7,900	4,000	25,800	40,700

[a]Includes cotton, rice, and peanuts in China; corn in France; and rice seed beds in the Orient.

[b]High tunnels are often computed together with plastic greenhouses in countries other than the Mediterranean. France is included in the Mediterranean group.

[c]Crude estimates only.

(Wittwer and Castilla, 1995). Crop maturity is hastened by 6 to 8 weeks. Water is the resource that now limits future expansion. There are near 3 million ha of plastic soil mulches for cotton, peanuts, rice, and corn. Plastic covers on high value row crops are now estimated at 250,000 ha worldwide, and soil mulches cover approximately 750,000 ha. (Wittwer and Castilla, 1994). Plastic covers, while having a significant impact on high value specialty horticultural crops and their increased availability, with improved quality, are still used on a relatively small scale in terms of overall global agricultural output.

Plastic structures over high value crops provide a useful housing containment for metering in extra CO_2 for growth enhancement. This has become a convenient practice for many such crops in North America and northern European countries (Wittwer, 1993). The use of plastic films for soil mulches, particularly in China, for staple crop production (rice, corn, cotton, peanuts) has become significant in combating climate variabilities. They have resulted in improved management of limited resources, and extended the boundaries, the quality, and magnitude of crop production. Such is optimization of global climate resources (Wittwer et al., 1987; Wittwer, 1993). The greatest potential for the future application of controlled environment technologies will be in upland agriculture, or what are now known in the developing world, as rain-fed areas, where water is constantly the limiting factor for plant growth.

The best plan for climate modification will be global reforestation. This would not only have a favorable effect on the albedo of areas so treated, but will, at least partially, mitigate the presumed threats that rising levels of CO_2 may impose (Dyson, 1992). Reforestation would also add to renewable resources, and greatly facilitate land stabilization in erosion-prone areas. Consideration should be given to the use of species that provide most of the wood. These include eucalyptus, radiata pine, Douglas fir, loblolly pine, black locust, Scotch pine, and aspen.

Given our present state of knowledge, with consideration of projected changes in global and regional climates and their consequences, there is no suggestion that, on a global scale, the supply of climate resources for food production will diminish or increase from present levels. However, the concomitant rising level of atmospheric CO_2 and other greenhouse gases poses another dimension.

There is no evidence that past climate changes, including those of the last decades, are related to changes in CO_2 *in the atmosphere, with the possible exception of warmer nights in certain parts of the Northern Hemisphere.* Our present knowledge of the climatic effects of the changing CO_2 content and other greenhouse gases in the atmosphere is totally inadequate as a basis for initiating any global attempt to change the climate (Bryson, 1993). However, the concomitant rising level of atmospheric CO_2 and other greenhouse gases poses another dimension.

Climate must be managed as a resource to be used wisely on the one hand, and a hazard to be dealt with on the other. One of the challenges for the future will be to combine improved climate information and new technologies for planning and decision making in the management of food, energy, and water resources.

3

Climatic Impacts from Rising Levels of Atmospheric Carbon Dioxide and Other Greenhouse Gases

3.1 INTRODUCTION: HISTORICAL RECORDS, CURRENT STATUS, AND PROJECTIONS

The magnitude of future impact of rising levels of CO_2 and other trace gases in the atmosphere on global food production has been, and remains, an area of great controversy. The greenhouse warming question is not whether atmospheric composition of trace gases is, or will be, a determinant of global average temperature but, rather, how large, how fast, and when and if temperatures may change as the gases increase in concentration. The earth's average temperature is now 15°C. The chemical climate is changing with increasing levels of water vapor, carbon dioxide (CO_2), methane (CH_4), nitrous oxide (N_2O), chlorofluorocarbons (CFCs), hydrogenated chlorofluorocarbons (HCFCs), and ozone (O_3). Without water vapor and trace gases in the atmosphere, it is believed that the earth's average temperature would be about 33°C colder than it now is (DOE Multi-Laboratory Committee Report, 1990; National Academy of Sciences, 1992).

The controversy centers first on the projections, now widely accepted by many climatologists and based primarily on general circulation models (GCMs). One group claims with a doubling of the preindustrial level of atmospheric CO_2, global average temperatures would increase between 1.9 and 5.2°C (3.4 and 9.4°F). The other group insists that such elevated temperature levels cannot be assumed to be in our future, because of the admitted unreliability and lack of validity of model projections. Further, the results from models, thus far, seem flawed in that real world observations concerning the already elevated levels of CO_2 (30%) and other gases are not reflected in current temperature changes. The qualitative comparison of the approximate 0.4°C rise in global average temperature over the past 100 years with the GCMs of greenhouse warming is problematic (Electric Power Research Institute, 1990). Alternative temperature increases of a much more modest level have been suggested (Seitz et al., 1989). Some suggest there may be little or no change (Balling, 1992; Lindzen, 1990; 1993; Michaels, 1993; Singer et al., 1991).

Second, is the issue of the many projected impacts of global warming on agriculture in general, and specifically on food production and food security, with the suggestion that many, but not all, resource impacts will be negative. Little, if any, credence is given to possible benefits of global warming and the many positive effects of elevated levels of atmospheric CO_2 on plant growth and biological output, which should accompany or precede, global warming.

Because of unpredictable factors, most forecasts miss reality by a large margin. This is especially true of those relating to population increases, the adequacy of food supplies and agricultural productivity, environmental degradation, and the exhaustion of resources. Some fail to recognize that some factors are unforeseen. It is impossible to know all possible inputs. The resource base can also change with time and technology.

3.2 TEMPERATURE

The potential of rising levels of atmospheric CO_2 producing a rise in surface global temperatures has been a topic of discussion since the idea was advanced nearly 100 years ago by Arrhenius (1896). GCMs have been, and remain, uncertain. They are the principal tools used by most, but not all, current forecasters of climate change. The adherents of the GCMs project that an increase in greenhouse gas concentrations (CO_2, CH_4, CFCs, O_3, N_2O), equivalent to a doubling of the preindustrial level of atmospheric CO_2, would produce global average temperature increases of between 1.9 and 5.2°C (3.4 and 9.4°F). The scariest of the scenerios is that the larger of these temperature increases would mean a climate that is warmer and with the increase produced more quickly than any known in human history. The consequences of such warming are not known, but many are quick to predict major dislocations in food production, and agriculture suffering with resultant widespread problems of food security.

Not all agree with such extreme levels of warming (Balling, 1992; Idso, 1989; Lindzen, 1993; Michaels, 1991; Michaels et al., 1993; Seitz et al., 1989a; Strommen, 1992) and project much more tolerable, and even favorable changes when coupled with the direct benefits of rising levels of atmospheric CO_2. They predict that there would be a net benefit for food production. The position of a very modest temperature change related to the lower levels of increase is supported by the record of the past 100 years, where the average global temperature has increased only between 0.3 and 0.6°C (0.5 to 1.1°F). During this time the CO_2 level has increased up to 30%, and has been added to by even more rapid increases in other so-called greenhouse gases. With these rather remarkable increases in atmospheric CO_2 it is difficult to explain such low increments in temperature change, when in reality with the increases in not only CO_2, but other greenhouse gases as well, we should already be approaching the halfway mark of temperature increases projected for a doubling of CO_2 by the GCMs. In other words, because of the combined effect of these so-called greenhouse gases, we have already gone over halfway to an equivalent doubling of CO_2 (Balling, 1992; Hansen et al., 1989; Michaels, 1990; Wigley, 1987). Even so, there has been less than a half a degree of warming in the past 100 years. This is not all. The minor temperature increases (0.3 to 0.6°C) that have occurred during the past century could be attributed either to greenhouse warming, heat island effects at weather stations, inadequate sampling, or natural climate variability. None can be ruled out.

Numerous other ranges of temperature increases, from theoretical and numerical analysis, have been projected, but the pattern is essentially the same. Taylor and MacCracken (1990) state that the projected warming arising from increasing levels of CO_2 and other atmospheric gases will range between 1.5 and 4.5°C sometime during the 21st century. This range of temperature is derived from some 28 predictions of GCMs that scatter from a fraction of 1°C to nearly 10°C (Balling, 1992; Idso, 1989a). These numbers show an amazingly large range of responses to a doubling of CO_2, and the span is proving hard to narrow except downward. This is because no one knows how feedbacks, such as changes in clouds, water vapor, sea ice, oceans, and unidentified terrestrial sinks, might amplify or dampen the CO_2 effect (Schneider, 1994).

The water vapor content in the atmosphere, although highly variable with time and season, is the most abundant greenhouse gas. It is only because water vapor is able to selectively trap outgoing longwave radiation that life on our earth is possible. Otherwise, it would be too cold to sustain life. Small increases in the earth's albedo relating to increased cloudiness expected from global warming will produce a net cooling effect far greater than the potential contribution of warming from all greenhouse gas increases including that of CO_2 (Strommen, 1992). It has been suggested that the world's clouds may have an effect on the earth's energy budget that is up to 10 times greater than would be produced by a doubling of the amount of CO_2 in the atmosphere (Wildavsky, 1992). A slight change in cloud cover could either negate global warming induced by greenhouse gases or greatly intensify it. It is concluded that until clouds can consistently find their way into GMCs, they will be of little value in such predictions (Abelson, 1990; Anonymous, 1993b). These GMC projections, in turn, have also been hotly contested and seem not to be supported thus far by real earth observations. The National Academy of Sciences (1991; 1992) stated that increases in atmospheric greenhouse gas concentrations will probably be followed by increases in the average global atmospheric temperature. But we cannot accurately predict how rapidly such changes may occur, or how intense they will be for any given atmospheric concentration or location. More important will be what regional changes in temperature, precipitation, wind speed, frost occurrence, and length of the growing season can be expected. Thus far, no large or rapid increases in temperature have occurred, and there is no evidence yet of a CO_2-induced temperature or climate change. One projection of the GCMs is that warming would start first and grow fastest at the poles. Yet a major study of satellite data has shown a slight cooling trend above the arctic, with warming only at night, and only in the southern hemisphere (Karl et al., 1993). Both the benefits of longer growing seasons and fewer killing frosts would have to be factored against liabilities of increased pest infestation. Decreased cropping areas would have to be considered if such conditions prevailed.

A series of very recent reports reveals that although the global average temperature has been increasing, perhaps by 0.5°C during the past 100 years, warming has been primarily due to nighttime rather than daytime temperature increases, appearing as a decrease in the day–night temperature differences over land.

During the past four decades this warming, mainly at night, has occurred over North America, most of Eurasia, and portions of Australia, Africa, and South America (Kukla et al., 1994). In contrast to most climate model expectations, no significant nighttime warming has been observed in the polar regions where most of it, according to climate models, was supposed to occur (Karl et al., 1993).

Plant growth, it is emphasized, is sensitive to both day and night temperatures independently, as well as to the differential between them. This phenomenon is called thermoperiodism. Two crucial phenological events, the initiation of flowering and seed, or fruit setting, are sensitive to night temperatures, at least for two crops, the soybean and the tomato. Warmer nights could either advance or delay flowering depending on geographic locations, with resultant decreases in yield. Likewise, seed germination could be impacted over a long time frame. An increase in night temperatures can influence crop yields by acting on a wide spectrum of metabolic processes (Mooney et al., 1994).

It seems very unlikely that current projections will come to pass. Our concern might well focus on a 1 or 2°C temperature change, which is well within the range of interannual variations.

There is one category of anthropogenic impact that has occurred over the past 150 years not entirely attributable to the rise in atmospheric greenhouse gases. It is inadvertent modification in urban areas or the so-called "heat island effect". These climate modifications relate not only to temperature but to summer rainfall increases, especially downwind of cities, cloud cover, air composition, visibility, fog and smog, aided by lower wind speeds. Sunshine is notably attenuated. Abatement efforts for some of the latter have been partially successful in some localities but not others. Some are virtually hopeless because of orographic settings. Notable are Mexico City, Los Angeles, and more recently the Salt Lake City area during some winter months. If the heat island effect is eliminated by noninclusion of weather stations in cities and airports, there has been no temperature change in the lower 48 states of the United States during the past 100 years. Although the United States is not one of the largest land areas of the earth, it is one portion of the earth's surface in which greenhouse warming is predicted by some GCMs to be most apparent. It also has one of the most sophisticated weather station records, and is a major food-producing country (Ruttan, 1990).

Changnon (1992) indicated that more than half of all North Americans live and work in anthropogenically generated urban climates that are drastically different than those of 100 to 150 years ago. Totally new climates have been created in cities and downwind from urban areas. Societies and individuals have successfully adapted to them. The magnitude and extremes of these climates in temperature and precipitation are at least similar to, if not greater than, those predicted by GCMs for the next 100 years. In fact, the average urban warming per 100 years ranges from 2 to 3°C in cities of the United States such as Boston, Cleveland, and Washington, D.C. According to Jones et al., (1990), global warming is inflated 2 to 20% by the heat island effect. Also, the urban-enhancing influence has been greatest in the increase of summer temperatures. Urban heat island effects have also been noted in Paris, Berlin, Vienna, London, Sao Paulo, Tokyo, and Shanghai. Changnon (1992) concludes that a study of the inadvertent weather and climate modification that has occurred over the past century can serve as a useful historical analog for analyzing many aspects of the global issue. The past does hold some guidance for the future. People in cities have adapted to these increases in temperature.

3.3 PRECIPITATION

The projected effects of a rising level of CO_2 and other greenhouse gases on precipitation distribution and intensity are even less definitive than those for temperature. Although there is near certainty and unanimous agreement among scientists that models are deficient in projecting temperature increases from a doubling of greenhouse gases when these projections range from 1.5 to 5°C, the uncertainty is even greater concerning water supplies. For the United States it is projected that some areas could be wetter than at present and others drier. Some computer simulations show drier summers in the interior of North America. The impact would be greatest in the arid West where small changes in precipitation produce a relatively large change overall. Global warming will accelerate the hydrologic cycle, increase water requirements for crops, and further deplete limited surface and groundwater reserves, and the areas where crops could be irrigated. Projections suggest that for the more humid East, water would continue to be abundant but irrigated areas would increase, The direct effects of CO_2, to be discussed in Chapter 4, however, could partially compensate for the increased requirement by reducing the amount of water that would escape through the leaves (Abelson, 1989a; Waggoner, 1990).

The annual crop losses from drought and inadequate or excessive precipitation exceed those of temperature aberrations (Boyer, 1982). History shows in the United States and elsewhere that there is a tapering-off of drought impacts and water shortages in both crops and people (Warrick and Bowden, 1981). Also minimum measured yields in the United States are higher (declined less) in recent droughts than for earlier ones for corn, wheat, and soybeans. Through technology, yields have trended upward through both good and bad years. Few farms have been abandoned or transferred during recent droughts. If there is a CO_2-induced decrease in precipitation, the effects are progressively being tempered by technology and changes in cultural practices.

Waggoner (1990), in a summary of effects of a CO_2 doubling, states that detailed analysis, taking into account the continents, oceans, and circulation of the atmosphere in a large computer model, suggests that global temperature will rise 2 to 5°C, there will be an increase of 5 to

15% in precipitation, a rise of 10 to 100 cm in sea levels, and that the interior of large continents and mid latitudes will be drier. There will be more precipitation in some places and less in others. Also, more CO_2 will make plants grow faster and temper transpiration losses of water. The 1992 IPCC Supplement (1993) prefers the temperature range of 1.5 to 4.5°C and a 2 to 4 cm per decade rise in sea level. Again, as with temperature projections, real world observations thus far do not confirm computer model projections.

3.4 RAPIDITY OF CLIMATE CHANGE

Most projections are based on temperature increases resulting from a doubling of the current or preindustrial atmospheric CO_2 levels. Doubling is expected to occur sometime during the middle or the latter half of the 21st century. The scenarios based on models show one thing and real world observations show another.

Even the projections at the lowest levels of temperature increases are above those that have transpired thus far in the real world and those that might be expected. Nevertheless, within a few centuries we are returning to the earth's atmosphere and oceans, the carbon stored in fossil fuels, organic matter, and sedimentary rocks, that has accumulated over millions of years. This has led one author to declare that the insatiable demands of modern society and the combustion of wood, coal, oil, and gas have literally changed the direction and speed of atmospheric chemical evolution, and that CO_2 is accumulating in the atmosphere at a rate thousands of times faster than normal (Benarde, 1992). This has led many other authors to project that we will experience within the next century a more progressive change in temperature and precipitation than at any time in recorded history (Schneider, 1990). Also that, since 1860, a global warming has already occurred of 0.5 to 0.7°C, which is both statistically significant and consistent with experience based on theory and models. Further, that the warming is rapid now and may become even more rapid as a result of warming itself, and this "rapid and continuous warming will not only be destructive to agriculture, but will also lead to widespread death of forest trees, uncertainty in water supplies and the flooding of coastal areas" (Houghton and Woodwell, 1989). It is further promulgated that global warming would diminish biological diversity by causing extinction of many species (Peters and Darling, 1985). Typical of many biased reports is that of Kellogg (1991b), with the implication that the majority of the scientific community involved in climate research are convinced of the reality of a current and future global warming and that a temperature rise of 0.6°C recorded by land stations and ships at sea is truly proof that the greenhouse effect is already here, as Hansen testified before U.S. Congress in the summer of 1988. It is further emphasized by Kellogg (1991a) that six tenths of a degree centigrade may not sound considerable, but is significant when the climate record for the past thousands of years is reconstructed. There does not appear to be any other time when such rapid change has occurred over a period of 100 years. When the earth came out of the last ice age, about 15,000 years ago, there was a larger change but that took many thousands of years to occur. It is often emphasized that in keeping with a warming trend, the 1980s appear to be the warmest decade on record during the past century, and 1988, 1987, and 1981 the warmest years, in that order (Schneider, 1989).

The reality of and the magnitude of temperature change during the past century have not received the concurrence of all, nor is there an agreement as to causes of change; finally serious doubt prevails, based upon the inadequacy of modeling, that temperature increases will be as large as earlier projected, and if they occur, they will be in the lower range of the oft quoted 1.5 to 4.5°C, when the atmospheric CO_2 level achieves a doubling of current levels. Jones and Wigley (1990) state succinctly that the causes of global warming are less certain than the trend itself. The uncertainty is even greater, as to changes in precipitation patterns, with the addition that a change in water availability would have greater impact than temperature changes of the order predicted (Peters and Darling, 1985).

The so-called rapidity of warming so frequently emphasized may not be as threatening for agriculture as often projected. The research usually cited to demonstrate the importance of rates of change has to do with the migration of forest trees. In the usual scenario of rapid warming, it is projected that trees will not be able to move as fast as the climate. Such a conclusion, as Ausubel (1991a) points out, must be assessed in the context of other changes, some climatic, likely to occur in a similar time frame. The heat islands around major cities of the world is one. A recent comparison of summer temperatures in Atlanta and a nearby rural weather station showed that Atlanta temperatures increased between 1974 and 1988 by 2°C. Also, each recent century has brought massive land and ecosystem transformations in the spread of agriculture and the growth of cities. One should separate the rate of change with the direction of change, and whether the climate change is expected, rather than how rapid it may be.

3.5 LENGTHENING OF GROWING SEASON

A global warming-induced variable for crop productivity, and it could be the most important for temperate zone agriculture, would be both an increase in the length and intensity of the growing season (Decker et al., 1985; Wittwer, 1980), or growing degree units (Hillel and Rosenzweig, 1989). Seasonal and interannual variabilities in rainfall and snow cover, length of the growing season, early and late frosts, soil temperatures, and thermal variability in growing degree days (heat sums) are the climatic concerns of farmers.

The length of the growing period is defined as the number of days when both temperature and moisture permit crop growth. Days with mean temperatures above 5°C and with soil moisture resulting from rainfall at least equivalent to half the potential evapotranspiration are considered favorable to growth (World Resources Institute, 1992–1993). With other factors not limiting for growth, temperatures in temperate or cool climate zones determine the length of the growing season. The effects are primarily on the timing of developmental processes and on rates of leaf expansion. Most processes of development and expansion (flowering, seed production, fruit and tuber growth, leaf expansion) can be described by a linear increase of rate with temperature, from a minimum threshold temperature, varying with crops, to an optimum, and then a linear decrease to a maximum limit above which processes cease (Monteith, 1981).

The temperature response for crop production depends on whether the growth is determinate as it is for cereal grains with discrete life cycles that end when the crop is mature, or indeterminate as in the potato. Here growth and yields continue as long as the temperature is above a minimum threshold. For determinate crops of temperate regions, higher temperatures decrease the life of the vegetative canopy, radiation interception, and potential yield. In contrast, higher temperatures increase the potential yield of indeterminate crops (Goudriaan and Unsworth, 1990). When higher temperatures are increased along with CO_2 concentrations, the two act in opposition according to a mechanistic model for winter wheat. If the CO_2 concentrations were doubled and daily weather from a typical year is used, grain yield was increased by 27%. When daily temperatures were increased by 3°C and CO_2 doubled, the more rapid development of the crop shortened the growing season, and the potential grain yield was only 15% greater, but maturity was shortened by 30 days. A combination of 4.5°C and doubled CO_2 resulted in the temperature and CO_2 factors almost canceling out. In some other determinate species, such as maize in northwestern

Europe, positive interaction of CO_2 and temperature was noted. A longer growing season is usually, but not always, profitable. In most agricultural regions of North America and in southern Europe, increased temperatures result in losses of crop yield. The highest yields of corn, wheat, and soybeans in the U.S. corn belt are under cool wet summers (Thompson, 1975).

Many questions can be raised as to the most effective length of the growing season. It is reasonable that a warmer climate will lead to an increase in the length of the frost-free period. It is suggested that raising the temperature by only 0.6°C would extend the frost-free growing season in the U.S. corn belt by 2 weeks. Conversely, lowering the average global temperature by less than 1°C would be associated with June frosts and earlier killing frosts in autumn, with widespread crop losses (Pimentel et al., 1992).

A recent report (Karl et al., 1993) pertinent to lengths of growing seasons states that although global average temperatures have been increasing, the warming has been primarily due to nighttime rather than daytime temperature increases, appearing as a decrease in the day–night temperature difference over land for the past 40 years. The potential benefits of nighttime warming, such as a longer growing season and fewer killing frosts, may be offset, however, by increased pest infestations, a reduced cropping area, and higher human heat-related mortality.

For practical temperate zone agriculture, the growing season begins in the spring when the soil is sufficiently dry for seed bed preparation, and warm enough for planted seeds to germinate. It ends with the first killing frost or freeze in the fall. However, cool and wet springtime weather, a frequent occurrence in the U.S. corn belt and elsewhere, can greatly delay field operations of land preparation, seeding, and cultivation. These questions concerning soil–water balance and warming of the soil in the spring have not been addressed by atmospheric modelers (Decker et al., 1985).

3.6 FROSTS AND FREEZES—FREQUENCY AND TIMING

Untimely frosts are a climatic hazard for crop production extending from the arctics to the tropics. They can cause widespread destruction of high value as well as staple crops in temperate zones in the spring and in the fall, and even during summer in the cooler regions. Wintertime freezing temperatures will periodically strike the winter fruit and vegetable crops in Texas, Florida, and California; the Mediterranean area of southern Europe; the Middle East, and the northern parts of Africa, with similar episodes in India and China; and in the southern hemisphere for Australia, New Zealand, Chili, Argentina, and South Africa. Climate variability, especially well recorded in recent years, has resulted in record low temperatures in many areas and disastrous freezes in subtropical areas. An example is a series of severe winter freezes in Florida, which damaged and killed citrus trees in 1962, 1971, 1977, 1981 to 1983, and again in 1985 and 1988. Such freezes set off the development of the Brazilian orange industry and a decline of Florida oranges in the world market during the 1980s (Waggoner, 1992). The freezes also caused a migration of citrus production southward in Florida to the more southern counties. Another example for a semi-tropical region was a frost in Brazil in 1975, which hit a major coffee region and left 600,000 people jobless. It would appear logical to conclude that the risk of frost and freezing temperature worldwide would be lessened under any global warming that might be induced by elevated levels of atmospheric CO_2 and other greenhouse gases.

3.7 FREQUENCY OF DROUGHT

There is now a large volume of literature on the impacts of drought on agriculture. One of the common projections from rising levels of atmospheric CO_2 is that the interior of the United States and other countries, such as Russia with large land masses, would be drier with greater prevalency of drought (Kellogg, 1991a). A question often posed is whether temperature and precipitation will become more variable with global warming. Research results, on this subject, to date, do not indicate an increase in variability (Ausubel, 1991a; Reifsnyder, 1989). If, however, the climate in some areas were to become drier on average the chances of extremely dry weather are likely to increase. Rosenberg (1988) cites the long-term drought in the Sahel of Africa through the 1970s and 1980s and that it suggests a true drought might be occurring there, and then counters that that drought is not unprecedented in either its severity or duration, and should not be taken as evidence of any global trend, either increasing or decreasing variability. White (1984) further states that drought disasters usually follow in the wake of prolonged periods of unusually dry weather in areas where land use and agricultural production are ill-prepared for such dry periods. The worst drought conditions in the United States, since the skies of the midwest were dust darkened in the 1930s, was in 1988. It was preceded, however, by dark days on 21 October 1716, 19 October 1762, 19 May 1780, 16 October 1785, and 3 July 1814 (Schultz, 1982). Even so, some parts of the nation were even wetter than normal. There is nothing to suggest greater frequency of drought, now or for the future.

3.8 ENHANCEMENT OF POTENTIAL EVAPOTRANSPIRATION

No definite trend has yet been identified in the United States or globally, for changes in either temperature or precipitation (Strommen, 1992). The effects of high CO_2 for increasing the photosynthetic rate and decreasing the transpiration rate will be thoroughly documented in the following chapters. The increases in photosynthesis are more pronounced in C_3 than in C_4 plants, while the decrease in transpiration is more pronounced in C_4 plants. As for transpirational losses, the increase in leaf area from higher CO_2 and higher leaf temperatures may compensate or offset, at least in part, decreases in stomatal openings and more extensive root growth (Allen et al., 1986). A warming of the world will directly speed up evaporation of moisture, because water vapor capacity of the air doubles for each 10°C increase in temperature. If global temperature rises with increasing CO_2, crop water use may actually increase. Any greenhouse warming, however, may alter cloudiness, wind movement, and humidity as well as temperature. If there is an increase in evaporation, or evapotranspiration, stream runoff will be reduced and there will be an increase in the demand for crop irrigation in areas now of marginal water supplies (Schneider, 1989). There are, however, projections that mean global precipitation may increase by 7 to 15% with a doubling of atmospheric CO_2 and other atmospheric trace gases.

Transpiration cools plants by the loss of heat to change water from a liquid to vapor. Partial closure of stomata restricts transpiration and causes leaf temperatures to rise. If there is no climate change, there should be no indirect impact on evapotranspiration from climate. CO_2-induced plant growth, however, will likely affect evapotranspiration through changes in plant size, leaf area, and root growth and possibly changes in stomatal behavior. Rosenberg et al., (1989) have, however, emphasized the hazards in assuming that global warming will increase evapotranspiration under all circumstances, and further that a CO_2 enrichment-induced stomatal closing will actually affect evapotranspiration in open fields over long periods of time and is truly persistent and not transitory. Regardless of what the evapotranspiration effects may be, rising global atmospheric CO_2 will increase photosynthesis, growth, productivity, and water-use efficiency of both C_3 and C_4 plants (Allen, 1991; Allen et al., 1986).

3.9 CLIMATE MODELING AND ANALOGS

Taylor and MacCracken (1990), in a detailed discussion of projections for climate changes, outline some of the advantages of three-dimensional GCMs over extrapolation from current trends, the sensitivity of the climate system to changes in radiative fluxes, temperature, and small scale perturbations that occur during a normal cycle of climate change. A more recent description of computer simulations of the global climate system, outlining limitations as well as potentials, has been prepared by the Atmospheric Environment Service of Canada (1994).

Another approach is that in which climate analogs to current situations are sought. With the climate analog approach, there is an attempt to identify a past climate situation comparable to the one now occurring. A common analog for the drought and heat wave in 1988 for the United States was the dust bowl years of the early 1930s. In other words, the data base of past climate history is used to predict future climates. The limitations of data, the incomplete understanding of causes of past climate changes, and the rapid increases in atmospheric CO_2 and other trace gas concentrations being experienced today preclude the identification of any suitable analog of the past for projecting the future.

Schneider (1989) in an oft quoted review entitled "Changing Climate" states that "There is virtually no doubt among atmospheric scientists, that increasing the concentration of carbon dioxide and other gases will increase the heat trapping and warm the climate." He goes on to state that reliance is on mathematical models at several U.S. laboratories and elsewhere. He further states the models are in rough agreement and that a doubling of atmospheric CO_2, or an equivalent increase in other trace gases will warm the earth's average surface temperature by 3.0 to 5.5°C. Further, he states that such a change would be unprecedented in human history, and it would equal the 5°C warming that occurred since the peak of the last ice age 18,000 years ago, but it would take effect 10 to 100 times faster. Furthermore, Schneider (1989) continues to discuss the negative aspects of a warming, coupled with changes in precipitation, that could threaten natural ecosystems, habitat quality, agricultural production, human settlements, and forest types, and reduce stream runoffs with water shortages.

The implication is that most scientists believe there will be dire consequences, irrespective of the fact that there is now a wide discrepancy between the predicted warming and what has been observed in the real world in temperature and moisture changes, following a near 30% increase in atmospheric CO_2 and comparable increases in other trace atmospheric gases now assessed to have a warming effect equal to that of CO_2. Simply stated, the warming has not occurred that is predicted by the GCMs. It is now clear that CO_2 models predicting a global warming, ranging from approximately 1.9 to 5.2°C, and greater by a factor of three or four at the poles (MacDonald et al., 1979), with a doubling of atmospheric CO_2, have not been validated by observational data. Furthermore, there seems little interest in doing so. The extreme lack of scientific credibility among some scientists is reflected in a statement by Schneider (1989): "To get some broad-based support, to capture the public's imagination,...we have to offer up scary scenarios, make simplified dramatic statements, and make little mention of any doubts we have.... Each of us has to decide what the right balance is between being effective and being honest."

Predictions of the effects of global warming on agriculture have not factored in the use of plant and crop models nor the now well-known effects of elevated levels of atmospheric CO_2 on increased photosynthetic and water use efficiencies. As indicated earlier, the 1992 IPCC supplement (1993) now prefers the temperature range 1.5 to 4.5°C rather than the 3.0 to 5.5°C as indicated by Schneider (1989). Reference is now being made to even a lower range of 1 to 4°C (Guilderson et al., 1994). Also, if warming occurs, it will likely be within the lower end of this range. General circulation models are even less adapted to estimating potential changes in precipitation than potential temperature change. Further, for a comparison of the estimated precipitation, the models output for key agricultural regions in the United States did not even agree on the sign of change (Strommen, 1992). Landsberg (1984) has stated that neither the picture of the climate system nor its representation by numerical models has been fully grasped by many who have drawn far-reaching conclusions from ongoing climate research. *Precipitation changes are the most important as far as agriculture is concerned. It is how much, how soon, and with what regional distribution it occurs.* Here the atmospheric scientists relying on their mathematical models are noticeably silent. As Landsberg (1984) has reported, 28 model experiments based on these groups of models scatter temperature changes from less than 1°C to nearly 10°C. Most cluster around 1.5 to 4.5°C above current levels, agreeing with the latest IPCC (1992) assessment of consensus. A substantial minority, approximating 25%, suggest 1°C or lower. We expect that future clustering will be around the lower range of the 1.5 to 4.5°C.

Jones and Wigley (1990) have further warned that models cannot prove the emissions of greenhouse gases will significantly alter the earth's climate, and that the causes of global warming, if it exists, are less certain than that trend itself. Although the earth has experienced an overall warming trend of half a degree since the late 19th century, there are many questions concerning reliability. The earlier averages were compiled from a much smaller set of stations comparable to those of today. Are biases related to urban warming (heat islands) removed? Finally, how well do land-based temperatures represent the climate of the earth, which is two thirds covered by water? Finally, in a recent report, when a doubling of atmospheric CO_2 was considered, the radiative forcing reported among 15 atmospheric general circulation models showed roughly a three fold variation in predicted increases in global mean surface temperature (Cess et al., 1993). *The sensitivity of the world's climate system to rising levels of atmospheric* CO_2 *has not been resolved* (Hansen et al., 1993).

Models have consistently predicted a greater warming in the northern than southern hemisphere because of the greater extent of land in the north, which responds to radiative forcing. The opposite, however, has occurred. There was a 0.3°C increase between 1955 and 1985 in the south with no warming in the north. Strangely, the IPCC supplementary report (1992) to the 1990 report states that "the size of the warming is broadly consistent with the predictions of climate models, but it is also of the same magnitude as natural climate variability." The IPCC supplementary report (1992) continues (emphasis is the author's): "There has been some clarifications of the nature of water vapor feedback, although the radiative effects of clouds and related process continue to be the major source of uncertainty and there remains uncertainties in the predicted changes in the upper tropospheric water vapor in the tropics. *Biological feedbacks have not been taken into account in simulations of climate change." Computer modeling of temperature for future climates say nothing about nonclimatic or direct effects.*

As reported by Wittwer (1994), an extensive review of the CO_2–climate issue by a joint committee of the World Meteorological Organization and International Council of Scientific Union (ICSU) reached a most perceptive conclusion in 1983, that still holds (UNEP, 1985; WMO, 1983). It was stated that by simple comparison of the overall temperature and CO_2 increase for the last 100 years, one must conclude that the climate (temperature) sensitivity to CO_2 is at the lower limit of mathematical climate computer model prediction. In other words, global temperatures in the past 100 years have risen less than half as much (0.3 to 0.7°C) as the same models, which now predict a warming of 1.5 to 4.5°C above current levels by the middle of the 21st century. This is based on the more than 25% rise in atmospheric CO_2 which has already occurred, along with major increases in other greenhouse gases. Furthermore, these simultaneous increases in greenhouse gases, other than CO_2, should now be the equivalent, as far as any warming is concerned, to the halfway point of doubling of CO_2, but this is not substantiated by model projections (Balling, 1992; Idso, 1989a).

3.10 CROP MODELING

Acock (1990), in a very perceptive review of climate models, concluded that GCMs used to make predictions about the future climates are flawed, and that they are not yet in a position to make impact assessments that are meaningful. Also, any results obtained by using climate or crop simulators in their current state of development are misleading. Acock (1990) concludes that only a few scientists are involved in developing crop simulators at present. He further suggests that areas of ignorance should be addressed rather than further refining our knowledge of familiar phenomenon. Acock further charges the agricultural research community, stating that "after years of agricultural research and millions of field experiments, we still cannot quantify completely the responses of plants to seasonal variations in light, temperature, water, nutrients, and to the interactions between them. Now we are adding CO_2, a variable that is expensive to control and normally involves the use of enclosures." The challenge is to build crop simulators or models that incorporate the various environmental factors that affect the basic plant processes. For example, there is a well-known three-way interaction on crop response between light, temperature, and CO_2. These complementary interactions will be discussed in Chapter 4. The ultimate question to ask is whether such efforts help us to predict climate change or climatic patterns, the most determinate factor in worldwide agricultural productivity, which should have a key spot in assessing priorities for agricultural research in the future. Some efforts in this direction are now in progress (Adams et al., 1990). Agricultural productivity is sensitive to climate and to any global climate change. Models from atmospheric science, plant science, and agricultural economics have been linked to explore this sensitivity, but the sensitivity and impact on economies strongly depend on the climate model used.

3.11 CLIMATE AND FOOD SCENARIOS

Landsberg (1984), in his review of global climatic trends, quoting a review by Hare (1983), states that *the only thing agreed upon is that atmospheric CO_2 is increasing. One can now add to this the documented increases in other greenhouse gases, methane, nitrous oxide, and the chlorofluorocarbons, and the reality of climate variability.* Furthermore, the pattern of increased levels of CO_2 and other greenhouse gases has been observed globally. This indicates the effectiveness of the earth's atmosphere in dispersing emissions from whatever the source, natural or anthropomorphic.

Out of the above simulations of climate for the future, resulting primarily from current versions of models that indicate that global surface air temperatures will rise by 1.5 to 4.5°C when the CO_2 concentration is doubled in the atmosphere, have come an array of scenarios, most of them seemingly designed to frighten the public. The effects of clouds, and the way that models treat them, of oceans, the sequestering of atmospheric CO_2 by plants, losses from the soil, volcanoes, and the direct effects of rising levels of CO_2 on plant photosynthesis, dry matter accumulation, and water use efficiency are not factored into such projections.

Kellogg and Schware (1981) have stated that the CO_2–climate problem, more correctly the CO_2–climate issue, is a cascade of uncertainty arising in part from a wide spectrum of scenarios. Those of Houghton and Woodwell (1989), for example, and those of Schneider (1989, 1992) predict rising sea levels inundating millions of hectares of land and coastal cities, of dislocations of agriculture, the desertification of what is now the United States breadbasket, of devastating droughts, changes in the locations of monsoons, disappearance of endangered species, and disruptions of forests and natural ecosystems, including the carbon balance of arctic ecosystems (Shaver et al., 1992), which may be unable to survive in their current localities. All these are well known, having been given great visibility in the press. Some scenarios, reviewed by Landsberg (1984), call for a drastic and rapid climate modification on a large scale with catastrophic possibilities never before witnessed in human history.

The most famous and widely referenced scenario is that of James Hansen, director of the Goddard Institute for Space Studies, who declared before the U.S. Congress during the peak of the drought and heat wave of the summer of 1988, that the odds were 99 to 1 that the greenhouse effect or global warming was already here. As supporting evidence, it is further emphasized that there has been about 0.5°C of unexplained "real" warming over the past 100 years, and that in keeping with the trend, the 1980's appear to be the warmest decade on record, with 1988, 1987, and 1981 the warmest years in that order. Although the inaccuracies of such statements and conclusions have been well documented (Balling, 1992, 1993; Strommen, 1992), the fallacy still exists. It has given way to widespread glamorized press releases, the publication of best sellers, an advocacy for policy decisions involving billions of dollars to restrict the burning of fossil fuels, and ammunition for many press releases, some in prestigious scientific journals, and the issuing of annual reports by Worldwatch, World Resources Institute, and other earth organizations detailing the apocalyptic path we are following. Embellishments of such reports, the clever use of caveats by Schneider (1989) and others, and the responsiveness of the press are discussed by Friedly (1993). *Frightening the public does get support for more research.*

Scenarios based upon the results of three-dimensional mathematical GCMs may be the best climatic predictive models available, but they are not well adapted for estimating potential changes in either global or regional temperatures or precipitation. No definite trends have yet been shown to exist in either. They are deficient in that they do not adequately incorporate the effects of clouds, which are expected to increase if there is a warming. As already indicated, clouds may have both negative and positive impacts on temperature. The negative effect comes from reflecting sunlight into space. The trapping of heat from below increases the temperature (Abelson, 1990). The water vapor content in the atmosphere, which is highly variable with time and season, is the atmosphere's most abundant greenhouse gas. A small increase in the earth's albedo related to increased cloudiness expected with global warming would produce a net cooling 5 to 10 times greater than the potential contribution for warming from all greenhouse gas increases (Strommen, 1992). When the water content and other characteristics of clouds were incorporated in one model, the predicted global warming from the doubling of CO_2 dropped from 5.2 to 1.9°C (Mitchell et al., 1990).

Still another very recent class of plant climate–CO_2 scenarios has taken over the press. It describes plant life in both the managed (agricultural) and natural ecosystems, including tropical rain forests and arctic tundra (Bazzaz and Fajer, 1992; Korner and Arnone, 1992). These scenarios are enriched with the words of uncertainty and ifs, but with negative implications of disaster and catastrophic outcomes.

Within a single report (Bazzaz and Fajer, 1992) are found the following "mays", some repeated several times: may, may be, may have, may also, may lead, may be associated, may be overstated, may slow down, may demonstrate, may enable, may undercut, may also be decreased, may complete, many forests may, and may induce. The "coulds" include: could allow, could enable, could undercut, could negate, could have, and could lower. There are further uncertainty enrichments with ifs, mights, will not necessarily, more likely than, some indeed, in some experiments, often, did little to improve, at first glance, but on closer examination, suggests, it appears that, there appear, can mean, and it is likely. Other recent reports (Polley et al., 1993) give a much more balanced predictive scenario of vegetation–climate–CO_2 impacts on biospheric carbon fixation, altered species abundances, and water use efficiency of biomass production.

Another scenario projects the impact of the several positive direct effects of the rising level of atmospheric CO_2 on plant growth and development, and resultant global food production, accompanied by a modest increase of one or two degrees or less in global atmospheric temperature. This scenario is based on the known facts that elevated levels of atmospheric CO_2 have the positive effect of substantially increasing photosynthetic efficiency, depressing both dark and photorespiration, increasing water use efficiency, root growth, and biological nitrogen fixation, reducing injury from air pollutants, and compensating, in part, for deficiencies in sunlight, water, adverse high and low temperatures, and soil nitrogen deficiencies [Council for Agricultural Science and Technology (CAST), 1992; Rogers et al., 1992a, b; Strain and Cure, 1985; Wittwer, 1985, 1990, 1992, 1994].

It is presumed that the beneficial effects of CO_2 on crop yields may offset some, if not all, of the adverse climate effects (Adams et al., 1990). Further, that overall biological production or biomass of the planet would increase or remain steady, but not decrease (National Academy of Science, 1992). It is further emphasized in this scenario that if there is a modest warming of a degree or two (CAST, 1992), this would likewise reflect the positive effects of a longer growing season and likely bring areas of both the northern and southern hemispheres into viable agriculture crop production, that are currently too cold for food or pasture crops.

Also, the boundaries of winter wheat production would be extended further north and replace much of the spring wheat in the United States, Canada, Russia, and China—all of which are major wheat producers. This would have a positive impact on food production because winter wheat is more productive than spring wheat. Overall, it has been conservatively estimated that global agricultural productivity has risen by approximately 2.5 to 10%, and possibly as high as 14% from the current increase in atmospheric CO_2 over preindustrial levels (Allen et al., 1987; Kimball, 1983a, b; Strain, 1992). It has also been estimated that the growing stock and forest growth in Europe has increased between 1971 and 1990 by 25 and 30%, respectively (Kauppi et al., 1992).

Finally, there is a scenario that offers that there has not been, and will likely not be a significant global warming, and if there is, it may be overall more desirable than the climate we now have. Also, that any climate change (warming) will be at the very lower levels of projections coming from the interpretation of projections of general circulation climate modeling. This scenario is based on the reality of what has happened in the earth's atmosphere thus far. It is emphasized that the computer projections of global warming are of known computational errors and do not include many variables, among which the most important is the effects of clouds, which may exceed by severalfold any anthropomorphic inputs of greenhouse gases, the inputs of oceans, and the biological sequestering by green plants and soils.

The rationale is that the models have predicted much more warming than has already occurred. The GCMs, with their known flaws, form the only basis for predicting catastrophic warming, from a doubling of greenhouse gases, and particularly CO_2, during the next century. If all the global warming of 0.4 to 0.6°C, which has occurred during the past century could be attributed to an increased greenhouse effect, it would still be only one-third to one-half of the lowest now being predicted by current models.

Most all such predictions are based on the climate change (temperature increase) that will occur when the atmospheric CO_2 content doubles. If one now adds the effects of other greenhouse gases, including methane, nitrous oxide and the chlorofluorocarbons, which have also increased, along with the 25 to 30% increase in CO_2 in the past 100 years, and have an approximate equivalent warming effect to that of CO_2, we have already gone halfway to an equivalent CO_2 doubling. This scenario is based on the following lines of evidence.

First, is the magnitude of observed warming. The observed record thus far for global warming is far below the average of 4.2°C (7.6°F) projected, and at the lowest levels of the ranges reported.

Second, are the timings and locations of the projected warming. The models have projected that most of the warming will occur in the high latitude winter, which presumes most of the warming to be at night. This would reduce or prevent deleterious temperature effects and would lengthen the growing seasons.

Finally, there would be the growth enhancement caused by CO_2. Carbon dioxide at its current atmospheric level, is a limiting nutrient for plants. A voluminous scientific literature has now demonstrated more growth through photosynthetic enhancement; improved CO_2 dark fixation, reductions in both photo and dark respiration; increased water use efficiency by plants, including food crops, and greater resistance to air pollutants, as the atmospheric CO_2 concentration increases. Except for the height of the ice ages, both global temperatures and CO_2 concentrations are currently near their lowest values for the past 100 million years. A prestigious group of scientists, including meteorologists and those familiar with the direct biological effects of elevated levels of atmospheric CO_2, is supportive of this scenario (Balling, 1992, 1993; Ellsaesser, 1993; Idso, 1989a; Landsberg, 1984; Lindzen, 1990; Michaels, 1992, 1993; Seitz et al., 1989; Reifsnyder, 1989).

4

Direct Effects of Rising Levels of Atmospheric Carbon Dioxide on Plants

4.1 INTRODUCTION AND HISTORY

Ordinary air now (1995) contains approximately 360 ppm of CO_2. To grow a crop of corn that will yield 6.7 Mg/ha (100 bushels per acre) requires 9 Mg (10 tons of CO_2). At the current atmospheric CO_2 concentration, corn plants must process over 30,000 tons of air to procure the needed 2.5 Mg (2.75 tons)/ha of carbon for the crop (Norman, 1962). (The average yield of corn in the United States was a record 138 bushels/acre in 1994.) It is a remarkable feat of biological engineering that enables terrestrial plants to appropriate approximately 15 billion tons of CO_2 a year from such a dilute medium as the earth's atmosphere. It would seem reasonable to assume that increasing the level of CO_2 in the atmosphere would be a welcome addition for the enhancement of overall plant growth, crop productivity, and total biological output.

Indeed, CO_2 fertilization of air has proven decidedly beneficial for crops grown in enclosed greenhouse structures, and also for plants and crops exposed to natural atmospheric conditions. Many experimental observations of the influence of elevated levels of atmospheric CO_2 on plants (Bazzaz, 1990; Bazzaz, et al., 1989; Dahlman, 1993; Eamus and Jarvis, 1989; Idso, 1989a; Kimball, 1983b; Rogers et al., 1981; Strain and Cure, 1986; Woodward, 1993; Wittwer, 1994) have found a number of repeatable effects across most species. These include, growth stimulation, especially of roots, reduced stomatal conduction and leaf nitrogen, and an increase in water use efficiency. Lawlor and Mitchell (1991) report, in a summary of field studies, that elevated atmospheric levels of CO_2 increase photosynthesis, dry matter production, and yield, substantially in C_3 species and less in C_4, decrease stomatal conductance and transpiration in C_3 and C_4 species, and greatly improve water use efficiency in all plants. Increased productivity is related to greater leaf area. Stimulation of yield is more from an increase in the number of yield-forming structures than their size. Partitioning of dry matter among organs is mostly evident in root and tuber crops. Over 1000 experiments have now been conducted around the world reporting favorable growth improvement or responses, approximating an overall increase in yield of 33% for all mature and immature plants from a doubling of the CO_2 concentration from its current level of 360 ppm (Kimball, 1985). These data have been further extrapolated by Goudriaan and Unsworth (1990), who suggested that about 5 to 10% of the actual rate of increase of agricultural productivity worldwide during the past century can be ascribed to the fertilizing effect of rising atmospheric CO_2. Rogers et al., (1994a) correctly state that "Carbon dioxide is the first molecular link from atmosphere to biosphere. It is essential for photosynthesis which sustains plant life, the basis of the entire food chain. No substance is more pivotal for ecosystems, either natural or managed." The rising levels of atmospheric CO_2 represent a resource for agriculture and food production rather than a conventional air pollutant.

The fertilizing effects of elevated levels of atmospheric CO_2 were noted and reported over 100 years ago, and probably observed much earlier. As early as 1888, the benefits of CO_2 were recognized and reported for practical greenhouse crop culture in Germany. Details of early experiments of almost 100 years ago, conducted in greenhouses in western and northern Europe, reporting significant increases in plant growth and productivity, have been summarized by Hicklenton (1988), Wittwer (1986), and Wittwer and Robb (1964).

Parallel to the extensive experiments with CO_2 enrichment for the enhancement of growth of crops grown within greenhouses, in growth chambers, or subject to other controlled environments, have been the studies on CO_2 fertilization relating to the basic photosynthetic process (Rabinowitch, 1956). They affirm that neither the normal CO_2 concentration of the air, nor the content of the gas in water equilibrated with the free atmosphere is sufficient for complete saturation of photosynthesis in either moderate or strong light. The curves indicate that it should be possible to improve, by as much as 50 to 100%, the yield of photosynthates under natural conditions by means of CO_2 fertilization, and that one may expect this to lead to a proportionate increase in crop yields. Also, experiments tend to confirm such a conclusion. There is also the basic question of how strongly the actual concentration of CO_2, immediately surrounding vegetation under natural conditions, deviates downward, from the average atmospheric content, or what that concentration should be for optimal yield response. This could be critical in nonventilated greenhouses, forest canopies, and in dense vegetation of field crops on a day with little or no air movement. These conditions often prevail under natural conditions. There is also the perennial question as to the extent to which the inherent low level of atmospheric CO_2 is a deterrent to plant growth and crop productivity, in general, if other factors such as sunlight, temperature, water, and soil nutrients are limited, on the one hand, and not limited on the other. For example, there is the current prevailing widespread presumption, often repeated in the scientific literature, that there can be little positive response to CO_2 fertilization if other factors, such as water, light, temperature, or soil nutrients, limit growth (Bazzaz and Fajer, 1992; Kramer, 1981; Strain, 1992). This erroneous premise will be discussed and reviewed later in some detail. New free air field plot designs and experiments are now in progress that should resolve this issue. There was also the early erroneous presumption that CO_2 enrichment may merely put more water into treated plants with no concomitant increase in dry matter content. In other words, there is more water packaged in green leaves. This presumption, promulgated by two major research reviews by Lemon (1977, 1983), was initially accepted by some, even though it was later shown to be erroneous.

Thus, we have now introduced two geophysical phenomena, occurring simultaneously, that could have and may now be having major consequences on world agriculture, food supplies, and future food security. These are the so-called indirect effect on global warming or the widely publicized "greenhouse effect", glamorized by some climate modelers, and environmental activists, and the direct stimulation of plant growth (Wittwer, 1988). We have thus far reviewed the so-called global warming in terms of its magnitude, as suggested by some (Intergovernmental Panel on Climate Change, 1992; Kellogg, 1991a, b; National Academy of Sciences, 1991; Schlesinger, 1993; Schneider, 1993) as well as the limitations that have been seriously challenged by others (Balling, 1992, 1993; Ellsaesser, 1993; Essex, 1986; Idso, 1980, 1984, 1985, 1989a; Lindzen, 1990, 1991, 1993; Michaels, 1993; Michaels and Stooksbury, 1992; Newell and Dopplick, 1979). This "greenhouse effect" is considered, by some, to cause major shifts in agricultural production areas and ranges of wild life species and ecosystems, death of forest trees, and disruption of food supplies, especially in developing (third world) countries (Houghton and Woodwell, 1989; Rosenzweig and Hillel, 1993b; Strain, 1992).

The second geophysical phenomenon is the stimulative effect of the rising atmospheric CO_2 enrichment on plant growth and productivity

(Acock and Allen, 1985; Allen, 1979; Allen et al., 1987; Cure and Acock, 1986; Idso, 1989a; Idso and Kimball, 1992; Kimball 1983b; Kimball et al., 1993a; Reifsnyder, 1989; Rosenberg, 1981; Wittwer, 1985). The effect is direct, is easily observed, has been subject to thousands of experimental trials, and is easily measured. Results now range from plants grown under the protection of greenhouses, and other controlled environment chambers, and open-top chambers (Leadley and Drake, 1993) for some of the major food crops of the world in open fields, forest tree species, aquatic plants, and those in wetlands and in the arctic tundra, to simulated tropical ecosystems. Finally, free air CO_2 experiments without any containment are now in progress for some of the world's major agricultural crops (Hendrey et al., 1993; Kimball, 1993b).

The open-top chamber and free air CO_2 experimental (FACE) methodologies are offering useful approaches to the studies of plant and crop responses to elevated levels of atmospheric CO_2. Open-top chambers, in use for over two decades, are still the most widely employed and most thoroughly studied experimental method for exposing field-grown plants to elevated CO_2 and other atmospheric gases and pollutants. As Leadley and Drake (1993) report, climate in these chambers tracks the dynamic changes in temperature, light, and rainfall experienced in open fields. There is a major drawback. The microenvironmental conditions of the test plants are artificial—mixing of outside air with the air in the chambers is not possible. There is not the control of CO_2 concentration possible with open field experiments. The FACE methodology has been viewed by some as a "real world" approach that may provide the best test yet devised for the effect of the rising CO_2 enrichment on natural and agricultural ecosystems and food crops (Hendrey et al., 1993). In the FACE approach CO_2 enrichment is achieved by applying a network of pipes near the ground in such a way as to provide elevated CO_2 concentrations to the ambient air of the plants. It is in an open-air setting in the field with no containment. Thus far the FACE system has been in use on cotton for four seasons with stable control of CO_2 at 550 ppm ±10%. Results with cotton showed that with CO_2-enriched plants there were significant increases in biomass accumulation both above, and especially below ground. Soil respiration was also increased in CO_2-enriched plots. The enriched plants matured earlier with greater agronomic yields. Water-use efficiency also increased with CO_2 enrichment as has been observed in many other studies. Studies are now in progress with wheat. Chief deterrents are cost, details of which are outlined by Hendrey et al. (1993), and elevated atmospheric CO_2 was not maintained continuously day and night, turned on and off diurnally with only a 14-h daily exposure thus not simulating real world conditions.

Results with all these different approaches to atmospheric CO_2 control generally have been positive. Of over a thousand experiments, detailed in 324 peer-reviewed scientific reports, 93% reported an increase in plant productivity averaging 52%. Only 2% gave a documented decrease (Idso, 1993).

These two on-going global experiments are under intense scrutiny (Adams et al., 1990; Council for Agricultural Sciences and Technology, 1992; Kimball, 1986b; Kimball et al., 1990a, b; Stitt, 1991; Strain and Cure, 1985, 1986). Little attention, however, has been given to their possible interactions. For example, as we have indicated, in addition to the enhancement of photosynthesis, higher levels of atmospheric CO_2 tend to partially close leaf stomata. This decreases transpiration per unit leaf area (Jones et al., 1985a, b) and raises leaf temperatures (Idso et al., 1986, 1987). Thus, such direct effects of CO_2 on temperature of plant parts or entire plants, coupled with possible indirect changes in potential air temperature, will be superimposed on the direct effects of CO_2 on plant growth and development. What are the consequences of temperature variability for CO_2 productivity relationships, and will increasing temperatures and possible changes in precipitation patterns augment or reduce the stimulation of plant growth and development with CO_2 enrichment (Allen et al., 1989b)?

There are important reasons for the increased crop productivity, dry matter increases, and higher yields from elevated levels of CO_2. First is the superior efficiency of photosynthesis, resulting in marked reductions of respiration. Net photosynthesis is the sum of gross photosynthesis minus respiration, including both photorespiration and dark respiration. The second is an increased water use efficiency resulting from sharp reductions in water loss per unit of leaf area. This relates to the partial closing of stomata associated with higher CO_2 levels. while stomata admit air and CO_2 into the leaf for photosynthesis, they are also the chief conduit for moisture loss. While higher CO_2 levels greatly reduce water loss, there is little if any impairment of CO_2 uptake. The mechanism by which the concentration of CO_2 in the atmosphere determines whether stomata tend to open or close is still largely unknown.

Other observed and possible benefits are substantial increases in root growth resulting in an increase in root/top ratios, reductions in injury from air pollutants, and a partial compensation for deficiencies of light, moisture, temperature, and some soil nutrients and greater resistance to soil salinity.

There are marked variations in responses among plant species, crop varieties, food crops, and forest trees, to elevated atmospheric levels of CO_2. These relate to differences in species, in the pathways for photosynthetic carbon fixation, respiratory losses, to the part harvested of economic value, to the plant type (determinate or interdeterminate), and to various photoperiodic, temperature, soil moisture, and soil nutrient regimes. Each of these determinants will be reviewed.

4.2 PHOTOSYNTHETIC CAPACITY

Plants are divided into three broad categories depending on differences in their photosynthetic pathways for the fixation of CO_2. They are labeled, academically, as C_3, C_4, and CAM (crassulaceous acid metabolism) plants.

Most green plants, fruit, nut, ornamental, shade, and forest trees and shrubs, algae, most cereal grains (rice wheat, oats, rye, barley), all legumes, roots, and tubers, and vegetable and fruit crops, comprising about 95% of the earth's plants, use the C_3 pathway. It is so named because the first products of photosynthesis have three carbon atoms per molecule. C_3 species have both qualitatively and quantitatively greater photosynthetic responses to elevated CO_2 levels than C_4 or CAM species. C_3 plants have the inherent ability for more photosynthesis if the atmospheric level of CO_2 is increased. This is because of the excess capacity for CO_2 reduction, which now is partially wasted in photorespiration (Acock and Allen, 1985; Allen, 1979; Bravdo, 1986; Gifford, 1974; Gifford et al., 1985; Pearcy and Bjorkman, 1983; Tolbert and Zelich, 1983).

A central feature of C_4 plants is their ability to raise internal leaf CO_2 concentrations above the CO_2 levels of the atmosphere or surrounding air. Thus, they can partially overcome the photosynthetically CO_2 limiting level of today's air. Two groups of plants can do this, the C_4 (corn, sugarcane, sorghum, millet, some grasses) and the CAM (pineapple, agave, cacti).

There are several consequences of agricultural significance. C_4 plants photosynthetically, and more completely, remove CO_2 from the air than C_3 plants. C_4 plants can more effectively avoid the stresses of high temperature and lack of water. C_4 plants have a competitive advantage when restricted CO_2 availability limits net photosynthesis, especially in hot semiarid conditions. C_4 plants have a lower compensation point than C_3 plants. The mechanism for concentrating CO_2 in the mesophyll and bundle sheath cells of the leaf allows C_4 plants to take advantage of the high light intensities of full sunlight at midday and the high temperatures of the tropics. It also means that atmospheric enhancement with higher levels of CO_2 may not benefit C_4 plants as much if light is limiting. However, with water stress or high temperatures, more CO_2 will be beneficial to C_4 species. The much higher affinity for CO_2 in C_4 plants permits them to maintain a more favorable ratio between net CO_2 uptake and stomatal conductance. To determine the comparative capacity of photosynthesis and crop productivity under different light, moisture regimes, soil nutrient levels, and, especially, temperatures in open fields with a higher atmospheric CO_2 remains the ultimate challenge (Tolbert and Zelitch, 1983). Finally, it should be emphasized that the only major food crops of any significance having C_4 photosynthetic metabolism, whether they be cereal, legume, sugar, roots, tubers, or vegetables, are corn, grain sorghum, pearl millet, and sugarcane. The only food crop of any significance having CAM photosynthetic metabolism is the pineapple.

As far as food and agricultural crops are concerned with variables thus far imposed, most growth responses to elevated levels of atmospheric CO_2 are favorable. They include increases in total dry weight, root growth, higher root/top ratios, leaf area, weight per unit area, leaf thickness, stem height, branching and seed, and fruit number and weight. Organ size may increase along with root/top ratios. The C/N ratio may also increase. Important for agriculture is that there is an increase in harvest index and the marketable product, and a shortening of the growing season with earlier maturity resulting in reductions in both water and pesticide requirement (Table 4.1).

Most of the experimental results thus far for an enhancement of photosynthetic efficiency have been related to a reduction, or near elimination, of photorespiration in C_3 plants. This, coupled with the enhancement of water use efficiency, has led to an approximate prediction of a 33% increase in the mean yield, with a 95% confidence level of 27 to 38% of most of the world's important C_3 and C_4 plants, resulting from an approximate doubling of the current atmospheric CO_2 concentration.

Poorter (1993) gives an updated review of the growth responses of 156 species of plants to elevated atmospheric CO_2 including crops, wild, and woody, C_4, and CAM species. Compared to C_3 species, the growth stimulating effect is less with C_4 plants, but certainly is not nil. The growth response or increases in dry weights from a doubling of current levels was 41% for C_3, 22% for C_4, and 15% for CAM plants. There may be exceptions. One CAM plant, *Opuntia fiscus-indica* (prickly pear cactus), has shown a 32% increase in dry weight and decrease in water conduction during both day and night (Cui and Nobel, 1994). Within C_3 species, the herbaceous crop plant species had a 58% increase, and the wild species 35%, the fast growing wild species 54%, and the slow growing species 23%. Species capable of symbiosis with nitrogen-fixing

TABLE 4.1 Means By Which the Rising Level of CO_2 Will Directly Reduce the Negative Impacts of Global Warming (Higher Temperatures and Water Shortages)

1. A significant increase in water use efficiency for all C_3 and C_4 crops occurs under a high CO_2 environment.
2. The beneficial effects of CO_2 enrichment on photosynthesis and plant vegetative growth of C_3 species are increased as air temperatures increase. Rice and the soybean, however, represent two possible exceptions.
3. Optimal temperatures for plant growth and photosynthesis increase. This is especially significant for arid zone crop species.
4. At low temperatures (below 15°C for many species), there is little or no response to CO_2 enrichment.
5. Root growth and root/top ratios are enhanced.
6. Greater resistance to salinity or alkalinity and drought.
7. Shortening of the growing season with earlier maturity and less water required.
8. Greater biological nitrogen fixation and root mycorrhization (fungal extension of roots) occurs, resulting in increased nutrient and water uptake.
9. The majority of the world's most noxious weeds are C_4 plants. Their growth in a higher CO_2 world will be less favored than crop plants and forest trees, most of which are C_3 plants.

organisms had a greater growth stimulation compared with other C_3 species. The common denominator for all three groups of more responsive C_3 plants may be their larger sink strength. Filling carbon sinks by CO_2 enrichment becomes especially specific in the grain filling of cereals, the pod filling of legumes, and the tuber and root filling stages of root crops (Stulen and den Hertog, 1993), and the fruit of some vegetable fruit crops (Nederhoff, 1994). Overdieck et al., (1988) showed, for example, that in the radish (*Raphanus sativus)* the hypocotyls provided a sink for assimilates. There was some tendency for herbaceous dicots to show a larger response than monocots. Thus, within the group of C_3 species, differences exist in the growth response to high CO_2. In almost all instances and with all species, high CO_2 concentrations do indeed stimulate plant growth (Table 4.1). This is essentially the same conclusion arrived at and so declared over 30 years ago following our first experiments with elevated levels of atmospheric CO_2 (Wittwer and Robb, 1964).

A classical exception often quoted is that of arctic tundra (Oechel and Strain, 1985). It, however, has been suggested that the dampened response may be from a low nutrient status (Billings et al., 1984), a decreased sensitivity at low temperatures (Idso et al., 1986, 1987), or both. A further explanation is offered by Johnson et al., (1993) in that a large CO_2 response may already be in process in light of the heightened sensitivity to recent CO_2 changes in the subambient range. The historical rise of atmospheric CO_2 from 170 to 360 ppm since the last glaciation may be having and has had profound effects on the vegetative dynamics of arctic ecosystems. Given, however, the hundreds of thousands of plant species presently on the earth, the task of determining their individual CO_2 and temperature responses is overwhelming. Fortunately for food production, we are concerned with only a few hundred, and more specifically, with only 25 to 30 that stand between the human population and starvation.

Most experiments on CO_2 enrichment of atmospheres for enhancement of plant growth, including many of those for natural ecosystems, have been maintained only during daylight hours. Included is a recent widely publicized report on artificial tropical ecosystems (Korner and Arnone, 1992). This may help explain why the plant growth response to atmospheric CO_2, in many experiments, is often less than a mirror image of the effect of elevated CO_2 concentrations on net photosynthesis.

There has been a lack of a positive relationship between observed net photosynthesis and plant growth rates (Idso and Kimball, 1991). For example, a downward regulation of photosynthetic capacity has been observed in some long-term CO_2 enrichment experiments with plants grown in pots (Idso and Kimball, 1991), and where root growth is restricted (Mousseau, 1993). Allen et al., (1988), however, found that for experiments in floating water with *Azolla pinnata,* plant growth with no physical restrictions closely mirrored induced changes in net photosynthesis. Thus, one possible explanation for the rapid loss of sensitivity to higher CO_2 is that other necessary growth factors, such as water, nutrients, light and temperature, in some instances, are more limiting than CO_2 itself in regulating growth (Johnson et al., 1993). There was a linear response in growth with that of net photosynthesis in the absence of restrictive growth. Indeed, a series of long-term open-top chamber experiments, very recently conducted and some still in progress by Idso and Kimball (1992, 1993) and Idso et al. (1991), have demonstrated that with citrus and three Australian tree species, and continuous day and night CO_2 enrichment, *the increased growth rates are linear with the enhancement of net photosynthesis as a result of elevated levels of atmospheric CO_2*. Not only was there a decrease in photorespiration at elevated CO_2 levels, but dark respiration dropped by approximately 50% for a 360 to 720 ppm doubling of the atmospheric CO_2 concentration. Meanwhile, net photosynthesis rose by a factor of two, and growth was 2.8 times greater at the higher CO_2 level. These experiments were conducted, and are still continuing (Graybill and Idso, 1993; Idso and Kimball, 1993), in open-air chambers over extended periods of time (3 to 5 years for sour orange), and with results now reported through the summer of 1991, for the Australian tree species. Garcia et al. (1994) state, however, that solitary trees give a substantially different response than those partially shaded by other plants. Isolated CO_2-enriched plants produce more foliage. This allows even more CO_2 to be fixed, substantially magnifying the growth enhancement produced by the increased photosynthetic efficiency of each unit area of CO_2-enriched leaf surface. This enhanced positive feedback, however, does not go uninhibited. With more foliage, there is more self-shading. New approaches to an accurate assessment of acclimatization or downward regulation of photosynthesis to growth of some species at elevated CO_2 concentrations have been reported by Long et al. (1993) and Besford (1993).

There is now mounting evidence, as reviewed by Baker and Allen (1993a, b), and supported by Mousseau (1993) for the horse chestnut, and several other plants species (Reuveni et al., 1993), that at least for some plant species, CO_2 enrichment in the dark, or continuous exposure, can directly inhibit respiration. Details as to possible mechanisms are given by Amthor (1991) and Baker et al. (1992c). One must conclude that only those experiments, that expose plants or ecosystems, natural or artificial, to continuous—both day and night—elevated levels of atmospheric CO_2 will be direct analogs and truly representative of real world atmospheric conditions when the current level of atmospheric CO_2 doubles or changes with any measurable increment through time. As Rogers et al. (1994b) have suggested, a redesign of some of the present day experimental protocols with continuous exposure to elevated atmospheric levels of CO_2 may be essential if we are to approach real world conditions.

The proposition that acclimation or a downward regulation of photosynthetic capacity induced by elevated levels of atmospheric CO_2 will largely negate the fertilization effect of higher than ambient levels of CO_2 on C_3 plants is not supported. While Stitt (1991) gives a number of mechanisms by which an increase in carbohydrates could produce a feedback inhibiting photosynthetic rates following CO_2 increases, Johnson et al. (1993) reported no significant photosynthetic acclimatizations evident for three C_3 species—oats, mesquite, and little bluestem—when grown from June to September in serpentine plant growth tunnels with CO_2 gradients.

The biochemical background for the differential responses of C_3, C_4, and CAM plants to elevated levels of atmospheric CO_2 reside in the following. Carbon dioxide from the atmosphere is fixed by all green plants through the action of the enzyme ribulose bisphosphate carboxylase/oxygenase (RuBPease) in the C_3 metabolic pathway. RuBPease is ubiquitous in photosynthetic plants, is the most abundant, and is one of the largest and most complex of enzymes. It, in addition to the C_3 pathway, also drives the CO_2 pathway or photorespiration cycle. The extent to which the enzyme behaves as a carboxylase or an oxygenase depends on the relative amounts of CO_2 and O_2 at the level of the chloroplast. Plants that have the additional dicarboxylic acid cycle (C_4) plus some CAM plants effectively concentrate CO_2 at the reaction site and isolate the RuBPease from the effect of changes in CO_2 concentration. C_3 plants, without these extra buffering pathways, respond both by an increase in the metabolic pathway and a reduction in the photorespiration cycle to CO_2 enrichment. Increasing the atmospheric CO_2 concentration increases the rate of carbon fixation greatly in C_3 plants and to a lesser extent in C_4 plants. As a group C_4 plants have an average photosynthetic rate about 50% higher than the C_3 plants. Increasing CO_2 levels should favor C_3 species by shifting any competitive balance in their direction (Johnson et al., 1993). This may be a useful working hypothesis for weed/crop relations. In their arid, natural habitats, CAM plants show little or no response to an increase in CO_2. With adequate water, however, they behave like C_3 plants. Excellent reviews of these phenomena are provided by Acock and Pasternak (1986), Beer (1986), Black (1986), Bravdo (1986), and Tolbert and Zelitch (1983). Black (1986) has noted that with CO_2 enrichment of CAM plants a maximum of 25% increase in growth can be expected and the CO_2 enrichment should be at night, but water use efficiency may be increased severalfold.

Of the thousands of experiments that have been conducted on the effects of elevated levels of atmospheric CO_2 on plant growth and behavior, most have been made only on the yield of harvestable parts (Acock and Pasternak, 1986). In addition, for most experiments, only two levels of atmospheric CO_2 have been maintained, ambient and enriched to between 600 and 1200 ppm. In fact, the majority have been ambient and a doubling of what was ambient. The ambient is progressively rising with each passing year. This change in ambient has become significant over the past two decades during which most experiments have been conducted. Most frequently, the ambient is uncontrolled and assumed to be the same as outside air. It is frequently much less, especially in greenhouses and plant growth chambers. Similarly, there is usually a range of CO_2 concentrations in what is experimentally increased or designated as being double of ambient.

There are other limitations in some research relating to the effects of elevated levels of atmospheric CO_2 on photosynthetic efficiency on crop yields and plant growth (Acock and Pasternak, 1986). Experiments have been done in artificially lit growth chambers with light intensities much lower and action spectra quite different from those of sunlight. Many are short term and have measured plants only during the seedling stage. This has been true of many tree species. Finally, in many experiments and continuing at this writing and with the most sophisticated equipment (Hendrey et al., 1993), elevated atmospheric CO_2 levels have been maintained only during the daylight hours. This could preclude impacts on dark respiration and is not characteristic of what has occurred and will occur in future decades in the real world of elevated levels of atmospheric CO_2. Higher levels will not be turned off during nighttime hours. In fact, CO_2 levels will most likely increase during the night and in the absence of photosynthetic fixation.

It has been observed with some plants that the increase in leaf and canopy photosynthesis will continue to increase up to 1000 ppm and more of CO_2 in the atmosphere. For rice, the optimal level for yield increase is between 1500 and 2000 ppm (Yoshida, 1973, 1976). Later reports confirm lower levels (IRRI, 1993). For unicellular algae, the optimal level is 10,000 to 50,000 ppm (Tolbert and Zelitch, 1983) and for greenhouse-grown crops it is 1000 ppm (Nederhoff, 1994). Levels of atmospheric CO_2 optimal for the growth of a given crop or plant species have seldom been used in experiments thus far conducted. The typical approach has been to double the current ambient level (Campbell et al., 1990). We do not know what the optimal levels should be for most crop species.

4.3 WATER USE EFFICIENCY (CO_2 UPTAKE RATE/ TRANSPIRATION RATE)

An important phenomenon associated with increasing levels of atmospheric CO_2 is that the leaf stomata of plants will partially close. This increases the resistance to transpiratory water loss, with decreases in leaf transpiration rates, and an increase in water use efficiency. This may also result in higher leaf and plant temperatures. The ratio of CO_2 taken up in photosynthesis to water lost in transpiration is "photosynthetic water use efficiency." Water use efficiency may also be expressed in terms of the amount of biomass gained for the amount of water lost in a given period of time (Acock, 1990).

Water stress is the single most limiting factor for worldwide food, rangeland, and forest production. When atmospheric CO_2 concentrations vary significantly, plant water relations are directly affected. Evidence now shows that improved water use efficiency at elevated levels of atmospheric CO_2 associated with what may be projected as a global warming is an important finding for agriculture, rangeland, forestry, and all natural ecosystems that are photosynthetically active. By far, the greater response resides with C_3 plants such as soybeans, wheat, potatoes, rice, cotton, and forest trees. On the other hand, stomatal conductance generally declines with increasing CO_2 concentrations in both C_3 and C_4 plants, perhaps favoring C_4 plants such as corn, sugarcane, and sorghum. Transpirational rates may not be reduced in direct proportion to reductions in stomatal conductance. The result may be an increase in leaf temperature, which increases the water vapor pressure inside the leaf. Pearcy and Bjorkman (1983) point out that in a mixed community of C_3 and C_4 plants, an increase in an atmospheric CO_2 concentration will likely benefit C_3 species the most, providing they can take advantage of the extra available water. Again for C_3 species, the major CO_2 effect is on assimilation rather than transpiration. For C_4 species, it is first for transpiration, and second for assimilation (Goudriaan and Unsworth, 1990).

Some potential benefits of elevated levels of atmospheric CO_2, which cause a simultaneous increase in photosynthetic rate and a decrease in transpiration rate on future crop productivity, were suggested by Downton et al. (1980) and Wong (1980) of Australia. They speculated that improved water use efficiency may counter the predicted climatic effects of increasing temperatures and reduced rainfall in important agricultural areas. Specifically, they suggested that the affinity of RuBPease for CO_2 as temperatures rise may shift the temperature optimum upward for photosynthesis. Improved water use efficiency under higher atmospheric CO_2 may enable plants to occupy habitats, which in the absence of such enrichment would be too arid to support life. Indeed, increases in root/shoot ratios, frequently observed under limiting conditions of water or nutrients from CO_2 enrichment, enables plants to explore a greater soil volume and acquire more water and nutrients (Stulen and den Hertog, 1993). Relative to the recent increase in CO_2 on vegetation is the change in natural vegetation now in progress. Grasslands throughout the world are being invaded by C_3 trees and shrubs. Although this phenomenon is being ascribed to the impact of people, it is now becoming evident that such explanations are not adequate. Repression of RuBPease synthesis by increased CO_2 in the atmosphere also has implications for future plant protein quality, since this enzyme is the most abundant of all leaf proteins, constituting up to 50% of the total in many C_3 plants.

That elevated levels of atmospheric CO_2 alleviate water stress in plants by increasing stomatal resistance has now been confirmed by many independent investigations for a large number of plant species

(Allen, 1991; Carlson and Bazzaz, 1980; Cure, 1985; Jones et al., 1985a, b; Kimball et al., 1990a, b, 1993a, b; Sionit et al., 1980, 1981; White, 1985; Woodward, 1993). Elevated levels of atmospheric CO_2 affect stomatal closure and thus result in a reduction in transpiration rates. These effects become progressively greater as the level of atmospheric CO_2 rises. But the reduction in stomatal resistance may not result in an equal reduction in evapotranspiration (Rosenberg et al., 1990). Although the effect on stomatal resistance may be reduced to half or less, it is not nullified. Yields of water-stressed wheat at high CO_2 levels were as large or larger than those from well watered wheat at normal CO_2 (Gifford, 1979a, b). Rogers et al. (1984) demonstrated that with soybeans, high atmospheric CO_2 not only promoted greater growth, but prevented the onset of severe water stress under conditions of low water availability. Idso (1989a) found that with five terrestrial and two aquatic plant species grown under 640 ppm of CO_2, compared to ambient (340 ppm), there was a growth enhancement of about 30%. As water became progressively less available, there was an additional increment of approximately 15% in dry weight from CO_2 enrichment. This resulted in a relative productivity advantage of severely stressed plants, of about 60%. Nevertheless, there was a greater absolute enhancement of growth with adequately watered plants.

In summary, C_3 plants photosynthesis may benefit in dry matter production from high atmospheric CO_2 in three ways. First, there is an enhancement of leaf expansion, second, an increase in the photosynthetic rate per unit leaf area, and finally, an increase in water use efficiency.

There are still many unknowns concerning the consumptive use of water in crop production as it may be affected by rising levels of atmospheric CO_2 and possible changes in temperature and precipitation patterns (Raschke, 1986; Waggoner, 1990). Speculations as to landscape-scale consequences are still premature in the absence of adequate field data for nonconfined plants of several species continuously exposed to several elevated levels of CO_2. As suggested by Acock and Allen (1985), plants are probably going to be larger and have a greater leaf area in the future higher CO_2 world. This will tend to increase transpiration water loss. Plants will also, to partially compensate, likely have a larger, more vigorous root system capable of extracting more water and nutrients from the soil (Rogers et al., 1992a, b). A CO_2-induced decrease in transpiration will make more thermal energy available because of increases in leaf temperatures for soil evaporation. Consequently, the amount of consumptive water use (and that required for irrigation) will be reduced by rising levels of atmospheric CO_2.

These reductions in water requirements will very likely be less than the potential 33 to 40% reduction in leaf transpiration that has been noted. Actual reductions in consumptive water use up to 10 to 20% might reasonably be expected (Goudriaan and Unsworth, 1990; Rosenberg et al., 1990). This could still be significant in global food production, where, aside from low atmospheric levels of CO_2, water is now the most limiting natural resource, and where the intensity of that shortage is projected to be progressively greater with time. Plants, whose relative stomatal closure at high CO_2 levels is more than their leaf area increase, are more likely to have a reduction in water use if warmer temperatures occur. As a global antitransparent in biological productivity, elevated levels of atmospheric CO_2 could reduce overall evaporative water loss and increase water availability for use in agriculture and industry (Riebsame, 1989). However, as many predict, along with a rise in global atmospheric CO_2 concentration, we may also see a partially compensatory rise in evapotranspiration (Martin et al., 1989), coupled with possible changes in global rainfall amounts and distribution (Bretherton et al., 1990; Mitchell et al., 1990).

Polley et al. (1993) ascribed a significant global alteration in biospheric carbon fixation and altered species abundance to the increase in water use efficiency and biomass production of C_3 plants, which are the bulk (95%) of the earth's vegetation, and to which all temperate zone forest trees belong. They grew three species each of C_3 and C_4 plants with daytime gradients of glacial to present atmospheric CO_2 concentrations (150–350 ppm) in 30-m-long chambers. Parameters related to leaf photosynthesis and water use efficiency from stable carbon isotope ratios of ^{13}C to ^{12}C. Whole leaves were used. Leaf water use efficiency and above-ground biomass per plant of C_3 species increased linearly and nearly proportionally with CO_2 concentrations. Their results indicated that water use efficiency of C_3 plants may have increased by 27% over the past 200 years, and 100% since the last glacial maximum. They further concluded that the increase in net assimilation or photosynthesis over stomatal conductance in C_3 plants must have extended the geographic range of some species into areas where precipitation was normally too low to support growth. A specific example could have been from the moist washes to the drier slope habitats in the Sonora desert of the southwest United States and Mexico. They cautioned against using present vegetation–climate relationships to reconstruct past climates from pollen or fossil records without incorporating potential direct effects of CO_2.

Although it has been repeatedly demonstrated that atmospheric CO_2 may largely compensate for significant deficiencies of soil moisture

(Gifford, 1979a, b; Goudriaan and Bijlsma, 1987; Polley et al., 1993; Sionit et al., 1980, 1981), the net consumptive use of water may not be greatly affected. Furthermore, although it has been calculated by Kimball and Idso (1983) that there is a reduction of 34% in transpiration with a doubling of atmospheric CO_2 concentrations by averaging the results of 46 observations and 8 plant species in short-term growth chamber experiments, the consumptive water use will not be proportionately reduced. This is because plants at higher CO_2 levels will likely be larger with greater leaf areas. This will tend to increase transpirational losses. Such plants, however, will likely compensate, in part, by more vigorous and larger root systems, and be able to extract more water from the soil (Table 4.1). Net reductions of up to 10% in consumptive water use, however, may be expected (Kimball, 1985). Consumptive water use in C_4 plants at high atmospheric levels of CO_2 will likely be reduced more than in C_3 plants. This is because the stomata of C_4 plants were responsive to elevated CO_2 concentrations. Yet photosynthesis is relatively unaffected. With C_4 and CAM plants, there may be very little photosynthetic enhancement from CO_2 enrichment until the plants are stressed for water. C_4 grasses showed little enhancement of growth except under low water treatments (Nie et al., 1992).

The above observations strongly suggest that a substantial research effort should be directed toward the direct effects of photosynthetic water use efficiency on plants with the rising level of atmospheric CO_2. The rise is real. The positive effects on plant growth, water use, and global food production are real. We should accelerate the current efforts in determining the short- and long-term effects of increments of atmospheric CO_2. These increments of atmospheric CO_2, to serve as true analogs for the real world of the future, should be maintained continually and over extended time sequences. This would be useful for global food production, the potential for water conservation, forestry output, and the total biological productivity of the earth.

For plants growing under dry conditions, enrichment of CO_2 in the atmosphere can become a partial substitute for water.

4.4 IRRIGATED CROP PRODUCTION

Projected changes in precipitation patterns and global temperatures associated with any global warming, and an increase in water use efficiencies resulting in a decrease in consumptive water use by plants, coupled with human population increases, will likely increase demands for irrigated crop production. If there is an increase in water demands prompted by rising temperatures and shifts or reductions in precipitation of the now agriculturally productive areas, it could mean, for the United States, increases in irrigated acreage in the Mississippi delta and northern Great Plains. This would be true, also, of the north China plain and for northwestern and much of central India and neighboring Pakistan. It would be true of the more humid areas of the globe that are not now irrigated but devoted to the production of staple grain crops and seed legumes. This trend is already occurring in the United States, China, India, and elsewhere. Corn and wheat belts will shift to areas that are now somewhat cooler and wetter (Decker et al., 1985). Paddy rice, because of its high water requirements, could be replaced in some areas by upland rice, corn, wheat, and soybeans, as is now occurring in some parts of Southeast Asia, China, and other oriental countries. The production of grain sorghum and millet, both C_4 plants, will take over what are now some of the drier wheat- and corn-producing areas.

For the United States, this will be the western parts of the corn and grain belt. For China, it could be the northern plains and the Yellow River valley. If there is, however, as Downton et al. (1980) have suggested, and as confirmed by Rosenberg et al. (1990) and others, a significant reduction in consumptive water use at increasingly higher levels of atmospheric CO_2 with no appreciable change in temperature, food production could be expanded into some arid and semiarid lands not now considered as having adequate water resources to support crop production. Such areas could include much of the western and southwestern United States, many parts of Russia, India, Pakistan, the North China plain, most Middle Eastern countries, southern Europe, especially Spain, North Africa, all the African countries of the Sahel, and many parts of Mexico, Chile, Brazil, and Australia.

4.5 INTERACTION WITH AIR POLLUTANTS

Reference was made in Chapter 2 to the chemical climate. Not only is the CO_2 in the atmosphere, along with other so-called greenhouse gases—methane, nitrous oxide, and chlorofluorocarbons—progressively increasing with each passing year, and with what may be global warming, but there are other gases and particulates as well. These are considered atmospheric pollutants, and include sulfur dioxide (SO_2), the nitrogen oxides (N_2O), and ozone (O_3). We have heard little about ozone, as a pollutant, during the past decade. With ozone, there has recently been a concern for a shortage rather than an excess. O_3, SO_2, N_2O, and H_2S have produced confirmed harmful effects on agricultural crop productivity, rangelands, and forests (Black, 1982; Heck, 1989; Heck et al., 1982, 1988; Heggestad et al., 1985). The manner in which these pollutants harm vegetation and disrupt physiological processes has been reviewed by Allen (1990), Darrall (1989), and Idso (1989a).

The above materials account for over 90% of all the reductions in agricultural productivity ascribed to air pollutants. Annual losses to food crop production are staggering and for the United States are estimated between $2 and 4 billion (Adams et al., 1984). One estimate sets a figure of $3 billion loss of crop production annually in the United States for ozone alone (Experiment Station Committee On Policy, 1994). There are also believed to be serious air pollutant losses in forests and in crop productivity in the industrialized areas of Japan, China, India, much of western and eastern Europe, and in the Nile River delta.

However, the effects of these atmospheric pollutants may enhance as well as detract from crop production. The SO_2 in the atmosphere has provided a source of needed sulfur in some agricultural areas, and for some major food crops with significant amounts being extracted from the air by the leaves of plants (Maugh, 1979), the result being that sulfur is now seldom added as a needed element in fertilizer application. Sufficient sulfur appears to be provided from the atmosphere, primarily through foliar uptake (Wittwer and Bukovac, 1969).

Concurrent with the rising level of atmospheric CO_2 is the burning of fossil fuels and the introduction of the other trace gases and particulates into the atmosphere. Emission rates reveal that about 1% of the gaseous effluents released are oxides of S and N (U.S. Environmental Protection Agency, 1977). With CO_2 emissions, both the direct biological and indirect climatic effects are long term and global in magnitude. For most pollutant gases, such as SO_2 and N_2O, the residence time is only a few days and the deposition patterns are localized. Thus, the effects of the emissions of SO_2, N_2O, and other pollutants have focused on immediate, and not long-term, crop and aquatic plants and terrestrial ecosystem stresses. The differences in time and space scales and modes of action have resulted in a, perhaps unfortunate, disassociation of research on effects of increasing atmospheric CO_2 from the effects of air pollutants. Origins are in large part from fossil fuels (Allen, 1990; Dahlman et al., 1986).

Four points are important with respect to agricultural and forest productivity and the above air pollutants. First, the effects are regional. Different regions of the United States, China, Europe, Egypt, Japan, Korea, Canada, and India may exhibit dissimilar effects in response to pollutants. Sulfur is a good example, where gases in the atmosphere may either prove a benefit or a disaster. Second is the subtle magnitude of the effects. Acid rain, arising largely from SO_2 and N_2O emissions, is a good example where only decadal effects can be observed (D'Itri, 1982). Third is the issue of multiple pollutants, which is often the situation under real world conditions. The fourth relates to the interaction of air pollutants with anthropogenic and environmental stresses, such as drought and biological stresses encountered from natural systems.

As global population increases, agricultural production intensifies, and industrialization expands, more toxic gaseous and other air pollutants can be expected to be injected into the atmosphere. Air quality standards in the United States and elsewhere, in large part, have addressed the needs of, and hazards to people, not green plants or vegetation. The concern for agriculture should be to identify accurately the sources of air pollutants, monitor changes, assess their effects on food production, and seek means of reducing sensitivity to them. One approach has been to identify chemicals that reduce the damage (Cathey and Heggestad, 1982). Another would be to review the affects of rising levels of atmospheric CO_2 on stomatal behavior, particularly partial closure. One could then note any reductions in stomatal conductance, associated with reducing the harmful effects of increasing ambient levels of atmospheric O_3, SO_2, and N_2O. The amelioration of the harmful affects of such air pollutants on plant productivity is either given only token treatment or completely ignored in some major reviews of the direct effects of rising levels of atmospheric CO_2 (Lemon, 1983; Rose, 1989; Waggoner, 1992).

Stomata provide the main entry points for air pollutants into leaves, but uptake from surface deposition can be significant, ranging from 10 to 25% of the total (Allen, 1990). Leaf cuticles are generally impermeable to most gaseous air pollutants. It seems, therefore, logical to presume that substances or processes that cause stomates to close might help protect plants from damage by air pollutants. The same mechanism that tends to close stomates and increase water use efficiency in plants when atmospheric levels of CO_2 rise acts as a protection against the air pollutants of SO_2, N_2O, and O_3 for both C_3 and some C_4 plants (Carlson and Bazzaz, 1982). This was first suggested by Rich (1963), and later confirmed by Majernik and Mansfield (1972). They showed that CO_2 enrichment (500 ppm) would cause partial closure of stomates in the presence of SO_2. Injury was reduced by 14 and 66% in pinto beans and tobacco, respectively. Similar results on alfalfa were reported by Hou et al. (1977). They also observed on alfalfa, a C_3 plant, that with a mixture of SO_2 (1 ppm) and N_2O (0.33 ppm), there was a reduction in net photosynthesis of 50% at ambient atmospheric CO_2 (315 ppm). At 645 ppm of CO_2 there was a reduction of only 25%. Actual growth of plants in the presence of the pollutant SO_2 (0.25 ppm) was reduced with C_3 plants at 300 ppm of CO_2, but not at 600 or 1200 ppm. Similar fumigation of C_4 plants with SO_2 did not reduce growth at the higher CO_2 concentration, but did at 300 ppm. It appears that higher atmospheric levels of CO_2 modify the detrimental effects of SO_2. Similar results have been recorded for oxides of nitrogen (Kimball, 1986a). Kimball further concludes that overall, it appears that CO_2 enrichment in a polluted atmosphere is beneficial, at least for C_3 plants normally grown in greenhouses. The benefit is twofold. First, CO_2 enrichment stimulates photosynthesis and, second, the CO_2 causes partial closure of stomates and restricts entry of the pollutant into the leaf. The protective effect appears to apply to most atmospheric pollutants including O_3, SO_2, and N_2O. The resultant growth rate, however, with CO_2 enrichment may not be as large in its presence as in its absence. Finally, the protection may not apply to all plants and it should not be regarded as a cure for all aerial pollution problems.

Much has been reported as to the effects of air pollution on woody plants and their vulnerability to pollutants. There are possible interactions between air pollutants and elevated levels of atmospheric CO_2 (Sionit and Kramer, 1986). Again, since increasing the concentration of CO_2 usually causes partial closure of stomates, it should reduce entry of pollutants into leaves. With Ponderosa pine, severely damaged by air pollutants, an elevated CO_2 concentration caused a significant increase in the rate of photosynthesis (Green and Wright, 1977; Allen, 1990). It has been further suggested by Sionit and Kramer (1986) that the increasing concentration of atmospheric CO_2 may result in long-term changes in species composition of forests. It has also been suggested that air pollution may do the same (Woodwell, 1970; Miller and McBride, 1975). It is impossible now to predict whether the damaging effects of air pollution and the more positive effects of increasing atmospheric CO_2 will be synergistic or antagonistic with respect to the survival of a given species.

Allen (1990), through elaborate calculations, concluded that for soybeans, the reduction in stomatal conductance by doubling the current ambient level of atmospheric CO_2 could potentially reduce the harmful effects of ambient O_3 and SO_2 by 15%. Information, however, on the interactions of CO_2 and air pollutants is limited because regional changes in air pollutants are occurring concurrently with global change in CO_2. Legumes, such as the soybean, a major food and industrial crop, are among the most vulnerable to damage by air pollutants. This may have important implications for both crop production on agricultural lands and plants in forest areas as well as for total global food output and biological productivity. Yield depressions have already been noted for current levels of the major atmospheric pollutants (O_3, SO_2, N_2O). The magnitude of annual crop losses from air pollutants can be staggering for some urban and industrialized areas.

It may be good fortune that the currently and artificially elevated levels of atmospheric CO_2 for crop production in greenhouses may alleviate, in part, some of the adverse effects of air pollutants now common in urban and industrialized areas, where many greenhouse-grown crops are produced. Also, some of the most intensified agricultural food-producing systems on earth are within or adjacent to urbanized and industrialized areas with highly polluted atmospheres in China, India, western and eastern Europe, Egypt, Japan, Taiwan, Korea, Indonesia, and the United States. As for a solution, Idso (1989a) has declared that one might be hard pressed to come up with a cheaper cure for air pollutant damages in crop productivity. Atmospheric CO_2 enrichment partially closes the door to air pollutant entry into plants. It also compensates for the increased difficulty that this phenomenon provides for its own special entry, because of the increased concentration gradient from the air to the leaf.

4.6 PEST CONTROL-SPECIES COMPETITION

Aside from the primary entry of air pollutants through stomata, is that of various plant pathogens. Rich (1963) noted stomata are the most important infection courts for the foliage pathogens that cannot penetrate the unbroken epidermis. If increasing atmospheric CO_2 concentrations progressively constrict plant stomatal apertures, the incidence of plant diseases may drop or at least, not increase.

The entire area of microbial effects on food crops and other higher plants, as related to the possibility of increases in organic carbon, together with limitations in nitrogen and/or phosphorus, will, according to Lamborg et al. (1983), result in a preferential increase in nitrogen fixation and mycorrhizal activities. This will likely occur as the expedient means for supplying required nutrients to sustain the predictive increases in primary productivity resulting from elevated levels of atmospheric CO_2. Little research has been conducted in this area.

Both the inorganic and organic quality and content of plant tissue may be altered under elevated levels of atmospheric CO_2 (Overdieck, 1993). The nature of change in plant composition will depend on both the environment and species involved. Oechel and Strain (1985), quoting

Strain and Bazzaz (1983), suggest that the likely outcome will be plant material with lower inorganic nutrient content, higher secondary chemical content, and higher usable energy levels. Relationships between plant nutrient content and herbivore (insect) injury have been noted. One of the early reports was from my observations as a graduate student at the University of Missouri at Columbia (Wittwer and Haseman, 1946). Thrips injury in spinach was markedly reduced and essentially absent in plants at high soil nitrogen levels. Plants at a low nitrogen level were destroyed. Strain and Cure (1985), in their Executive Summary of the widely publicized "Direct Effects of Increasing Carbon Dioxide on Vegetation", state that soybean leaves become carbon rich and nitrogen poor as atmospheric CO_2 is increased, without the addition of more soil nitrogen. They then report that an insect pest, the soybean looper, feeding on the biochemically changed leaves had to consume more leaf tissue to gain an equal amount of protein nitrogen. Their broad conclusion from this experiment was that although agricultural productivity may increase in the future, there also may be larger increases of insect feeding rates and weed growth, and this would adversely affect the difficulty and expense of pest control. Strain and Cure (1985) conclude in their that not all species of plants respond to CO_2 in the same manner. They correctly state that plants with high conductance for the diffusion of CO_2 (presumably C_3 plants) will have greater growth than plants with lower CO_2 conductance. They then conclude that because many agricultural weeds have high conductance, they have a comparatively larger growth response to increased CO_2 than some desirable crop species that have lower conductances. The facts are that of the 20 most important food crops of the world, 16 have a high CO_2 conductance (C_3 plants), and of the world's 18 most noxious weeds, 14 have a low CO_2 conductance (C_4 plants), and 19 of the 38 major weeds of corn in the United States are C_3 plants (Wittwer, 1985).

Further review of research on pest control and species competition relating to elevated levels of atmospheric CO_2 gives no clear view of a future higher CO_2 world and agricultural pest problems assessing the effects on plants and their herbivores. Oechel and Strain (1985) quote again the work of Lincoln et al. (1984), where soybean looper larvae were fed soybean leaves grown at 650 ppm of CO_2 and consumed 80% more leaf material than larvae consuming leaves grown at 350 ppm. They correlated the greater herbivore feeding with the lower leaf nitrogen content. The suggestion was that the greater plant productivity at elevated CO_2 may be offset, in part, by increased herbivory. Overdieck et al. (1988) noted in a model pasture ecosystem that plant lice infestations broke out at elevated CO_2 levels. Strain added a personal observation (Oechel and Strain, 1985). He noted in the Duke University Phytotron that plants in elevated CO_2 chambers were more heavily infested with aphids than those at the contempory level of CO_2. Interestingly, Oechel and Strain (1985) then conclude, from their very limited sampling, that "most observations to date indicate that plants grown at elevated CO_2 will be more prone to insect infestation, and, once infected, the pests may consume more material than they would if the plant had been grown at ambient CO_2."

Not all agree with this broad conclusion. Kimball and associates (Butler et al., 1986) found no differences with four generations of sweet potato white fly or thrips (Butler, 1985) on developing cotton plants grown at elevated levels of CO_2 in open-top field chambers. However, populations of flea beetles, leaf hoppers, predaceous flies, and pink bollworms were much reduced at the higher (500 and 650 ppm) CO_2 treatment, compared to those at the ambient level. Beet army worms feeding on cotton grown at 640 ppm, compared to 320 ppm of atmospheric CO_2, experienced significant increases in times of development and mortality, as well as significant decreases in growth (Akey and Kimball, 1989). Plant-eating caterpillars performed poorly on a diet of CO_2-enriched foliage compared to that grown at the ambient level (Akey et al., 1988; Lincoln et al., 1986; Osbrink et al., 1987). The population of whiteflies was significantly reduced in tomatoes in CO_2-enriched greenhouses (Tripp et al., 1992).

Detailed studies of insect foraging or herbivory by Fajer et al. (1989) reveal that early instar larvae of the buckeye butterfly (*Junonia coenia*) fed on plantain foliage grown at CO_2 atmospheres of 700 ppm grew more slowly and suffered nearly three times the mortality of larvae raised on ambient CO_2 foliage. With late instar larvae, the results were quite different. They compensated for the reduced nitrogen content of the foliage by consuming more of it. Moreover, the larvae did not develop as rapidly and died more frequently. Slower growth could mean fewer reach adulthood because the caterpillars remain vulnerable to attack from predators and parasites for a longer time. Also, fewer caterpillars may complete development before dry or cold seasons set in. Consequently, it is likely that populations will decline. Bazzaz and Fajer (1992) continue with the argument that if herbivores suffer population reductions in a world of high CO_2, many predators would have less prey, and some predators, for example, feast on other insects that damage certain crops. Clearly, the jury is still out.

Idso (1989a), with a thorough review of the literature then available, suggests a tipping of the scales in favor of plants over foraging insects in a higher CO_2 world. This relates to the assumption that insect damage and attack may be partially or totally circumvented by the more rapid vegetative growth and earlier maturity of plants grown at elevated levels of atmospheric CO_2. Earlier maturity induced by elevated levels of atmospheric CO_2 should enable some crops to escape the ravages of insect invasions, as well as disease and weed infestations, and drought. The Chinese have been very adept at this approach. In the United States, this has already been partially achieved by genetically selecting cotton, corn, wheat, and soybeans for early maturity. Clearly, more laboratory and field experiments are required. It is unfortunate that the large-scale free air CO_2 experiments (FACE) underway with cotton and wheat include neither predator nor pollinator interactions, or even more importantly, competition with other crop species, and weeds (Kimball et al., 1993b). Stinner et al. (1989), with primary emphasis on global warming associated with rising levels of atmospheric CO_2 and pest problems, give little mention of plant growth enhancement, and reported only detrimental effects.

Thompson and Drake (1994) report for both a C_3 sedge and a C_4 grass exposed to elevated CO_2 in open top chambers in the field, there was less infestation of insects, and the severity of infection from pathogenic fungi was reduced. For insects the number of plants infested and tissue consumed were decreased.

Lincoln (1993) summarizes a number of studies, including his own, on the influence of plant CO_2 levels and nutrient supply on susceptibility to insect herbivores. Although competitive dominance in plants may be altered by elevated levels of atmospheric CO_2, and plant community structure changed, so leaf-eating herbivores may also strongly influence plant productivity. Increased CO_2 for plants may lead to increased feeding by lepidopteran caterpillars and grasshoppers, since leaf nitrogen is a limiting nutrient in many insect diets. Both generalist and specialist feeding insects respond to reduced nitrogen quality of their host leaves by increasing consumption rates. On the other hand, CO_2 enrichment may increase leaf fiber content, which could reduce leaf digestibility to herbivores. Leaf water content may either increase or decrease when host plants are grown in an enriched CO_2 environment and this could be an important factor influencing herbivore feeding. Lincoln (1993) concludes that it is apparent that plant/herbivore interactions can be anticipated in a higher CO_2 world. An altered nutritional quality of leaves appears as a generalized response, at least for C_3-pathway plants. There is, however, as some have speculated, little direct evidence, as yet, that leaf defensive phenolic type allelochemical content will be altered directly by CO_2 fertilization. However, there may be indirect effects if the increased production of plants at elevated CO_2 is not matched by similarly increased nutrient uptake. There is the suggestion that slow growing species of C_3 plants in contrast to fast growing ones with a large capacity to accumulate phenolic defensive compounds may have less herbivory resulting from the greater presence of such materials in a CO_2-rich world. Since C_4 plants show little or no response to elevated atmospheric CO_2 in terms of leaf nitrogen content (Curtis et al., 1989), the expected competitive advantage of C_3 plants, relative to C_4 plants, at elevated levels of atmospheric CO_2 may be diminished (Lambers, 1993). The certain outcome of a climate change is not more or less plant pests, but rather different ones (National Academy of Sciences, 1992).

Insects and diseases of both crops and livestock will surely migrate into new agricultural territories with any warming trend and/or changes in precipitation patterns accompanying rising levels of atmospheric CO_2 and other greenhouse gases. In the United States and Europe, the number of generations of European corn borer would increase and the devastation of the corn earworm would extend northward. Soybean insects and diseases would reduce yields in the northern states of the United States and in the provinces of China, as they now do in the southern parts. In Brazil and Argentina predators would move southward. The Xinjiang autonomous region of China, now relatively free of rice insects, would likely be invaded with a warming trend. Higher precipitation and humidity, which may accompany any global warming, would favor fungal diseases that attack cereals, particularly on wheat that cannot now be grown in the humid tropics. Many livestock diseases and insect vectors that transmit them, now limited to tropical regions, may spread in the future. Rift Valley fever, African swine fever, and the tsetse fly, the carrier of sleeping sickness, are good examples (Stem et al., 1989), along with horn flies in the United States and cattle ticks in Australia (Nilsson, 1992; Schmidtmann and Miller, 1989).

With the rising importance of integrated pest management, coupled with the need for sustainable food and agricultural production and the concomitant emphasis on cultural practices, genetic resistance, and biological controls as alternatives to chemical pesticide usage, the impact of the ever rising levels of atmospheric CO_2, with or without global warming, on plant–predator interactions, needs to be known. In both developed and less developed countries, after drought farmers most fear the uncontrollable devastations of insect pests, plant diseases, and weeds.

Plant competitiveness, particularly crops versus weeds, poses an interesting dilemma for evaluating the effects of the rising level of atmospheric CO_2 on global food production. Differential growth responses to both changing CO_2 concentrations and climate change will affect the future competitiveness and welfare of plants. There is much more agreement, however, on the direct physiological effects of increasing CO_2 on plants than on the indirect effects of CO_2 on global climate. This relates to the relative importance of various weed species in agroecosystems (Patterson and Flint, 1990) where weeds and crops of both C_3 and C_4 metabolism coexist and compete. Crops in agricultural fields, contrary to widespread opinions, are seldom, if ever, grown as true monocultures.

Weeds compete directly with crops for the limited resources of water, sunlight, mineral nutrients, and atmospheric CO_2 on which both depend for plant growth. This competition is costly. Of all crop pests, weeds are the most damaging. Annual losses in crop yields and quality for the United States' farmer from weeds alone are estimated at 10% of total production, or approximating $18 billion (Patterson and Flint, 1990; Waggoner, 1983). Percentages and total losses, for many developing countries, are even higher when the direct losses, and that which is used, including human labor, for weed control are considered. There are also serious environmental consequences arising from continued use of chemical herbicides, the most widely used pesticides, which will not be alleviated by the development of herbicide-resistant crops. It would be good fortune, if weeds were to do more poorly than crops in the approaching high CO_2 world. With agricultural-producing systems, one might expect that C_3 crops would generally fare better than C_4 crops, when the atmospheric CO_2 content is increased (Patterson and Flint, 1980; Patterson et al., 1984; Patterson, 1986). C_4 weeds in the presence of C_3 crops are the most common. Of the 20 food crops most important for feeding the world's population 16 have a C_3 photosynthetic pathway (Wittwer, 1981a). The only exceptions are corn, sorghum, millet, and sugarcane, which have C_4 photosynthetic pathways. Conversely, for few C_4 plants many of the major weeds are C_3 plants (Wittwer, 1985). As indicated earlier, 19 of the 38 major weeds of corn, the most important crop in the United States, are C_3 plants (USDA, 1972).

It is now well known that C_3 and C_4 plants respond differently to atmospheric CO_2 enrichment. These widely reported differential responses should be particularly relevant to weed/crop competition for agriculture in a future high CO_2 world. Early controlled environment experiments with plants grown in pots demonstrated that C_3 species would produce more biomass than C_4 species when grown individually under enhanced CO_2. In competition with plants of the two photosynthetic pathways at high CO_2, the C_3 plant performance was even greater (Patterson and Flint, 1990). Meanwhile, many additional experiments with C_3 and C_4 plants dealing with crop/weed relationships have been conducted. There are instances where the effects of CO_2 enrichment, other than the direct enhancement of photosynthesis, may favor C_4 rather than C_3 species (Bazzaz et al., 1989). It has also been observed that the competitive ability of C_3 species improved in assemblages of C_3 and C_4 plants in enriched CO_2 atmospheres, compared with ambient levels. Total community production also increased. An overall survey of current reports, when multiple species competition is considered, provides evidence that CO_2 enrichment is likely to increase the relative competitive ability of C_3 plants with respect to C_4 plants (Patterson and Flint, 1990). If a significant global warming, however, occurs with rising levels of atmospheric CO_2, and this is accompanied by increasing regional aridity as is predicted for some climate scenarios and for some important crop-producing species, the greater benefit may occur with C_4 plants. Any significant global warming may also have effects on noxious tropical or semitropical weed migrations northward in the United States and elsewhere, now limited by unfavorable low temperatures. Specific examples are itch grass (*Rottboelllia exaltata* L.) (Patterson and Flint, 1979), Japanese Honeysuckle (*Lonicera japonica,* Thumb), and Kudzu (*Pueraria lobata.* Wald.) (Sasek and Strain, 1990).

Weed/crop ratios for significant growth parameters have been altered significantly (Patterson and Flint, 1990; Wittwer, 1985). Any increase in global atmospheric CO_2 concentrations and/or temperature will influence weed/crop interactions and relationships. For agriculture, the influence is primarily with cultivated crops, but with far-reaching implications. Globally, all natural ecosystems will be impacted. The agricultural influence will vary with the weed/crop species involved, the photosynthetic pathway, growth habit, seed size, sunlight, photoperiod, soil–mineral nutrient level, water supply, human efforts to control weeds, changes in leaf anatomy, and surface characteristics induced by high CO_2. Increased rhizome, tuber, and root growth in difficult-to-control perennial weeds, many of C_4 metabolism, could be a likely consequence under high CO_2. *Amaranthus retroflexus,* more commonly known as "pigweed," is one of the world's most noxious weeds. It is also a C_4 plant but shows a remarkable stimulation to elevated levels of atmospheric CO_2 (Follett, 1993). Some strains have also become resistant to triazine herbicides (Zangerl and Bazzaz, 1984). The now demonstrated differential effects of CO_2 enrichment on the growth of C_3, C_4, and CAM plants under conditions of limited water, adverse (high?) temperatures, air pollutants, restricted sunlight, and limited soil nutrients are not known (Patterson, 1982). This is a vast area for future research. These environmental constraints, single or in combination, are those most frequently encountered in the real world of food and agricultural production.

For the future, it is conceivable the weeds, having a C_3 photosynthetic pathway, could, at higher levels of atmospheric CO_2 and no appreciable change in temperature, become more competitive with C_4 food plants, such as corn, sorghum, sugarcane, and millet. With a global warming accompanying elevated levels of atmospheric CO_2, the reverse could occur. Conversely, cotton, soybeans, cowpeas, field beans, mung beans, peanuts, wheat, rice, potatoes, sweet potatoes, cassava, sugar beets, bananas, coconuts, and most all horticultural crops and forest trees are C_3 plants and are now plagued by weeds, particularly the grasses, many of them C_4 plants. These important agricultural crops may become better competitors with weeds in a CO_2-enriched world (Table 4.1). With an accompanying global warming of some magnitude, the reverse could occur. Kimball (1985) has stated it well: "Even as C_4 species face stiffer direct weed competition, so will they face greater competition from C_3 crops for a share of the market."

4.7 CROP RESPONSES–DIFFERENTIAL EFFECTS ON PLANT PARTS

Counterpart to evidence of a significant CO_2-induced global warming trend, which is still uncertain after more than a decade of controversy, and seeking a first detection (Wittwer, 1981b), is accumulating evidence of the greening of the earth, especially in the agricultural and forestry areas. The evidence comes from many sources. Globally, agricultural crop yields are projected to increase by an average of 33% (Kimball, 1983a, b), with a doubling of current ambient atmospheric CO_2 concentrations. This has led others to suggest that overall crops yields have already increased by a conservative 2.5 to 10% (Strain, 1992) from the current ambient level of 360 ppm, which is estimated as 30% higher than the preindustrial level of about 270 ppm. Atmospheric levels of CO_2 are still rising. Allen et al. (1987) suggested in 1985 that soybean yields may have increased by 13% from 1800 AD because of global CO_2 increases. The so called "green revolution" of the 1960s to 1980s has coincided with the most rapid measurable rises in atmospheric CO_2. A reasonable current estimate would be an increase ranging from 5 to 10% for crop production (Goudriaan and Unsworth, 1990). It has also been suggested that from a further increase of CO_2 in the atmosphere to 400 ppm, the level projected by some for the year 2020 would probably result in a 20% increase in photosynthetic rates for some plants, provided other growth factors were not limited (Allen, 1979). The response of forest trees may be even greater according to a recent report from Finland (Kauppi et al., 1992). A 25 to 30% increase in the growing stock of forests in western Europe between 1971 and 1990 is attributed in part, to a 9% increase in atmospheric CO_2 during the same period. In very recent studies by Idso and Kimball (1992, 1993), citrus exposed continuously to two to three times the ambient level of CO_2 showed increased growth enhancement by a factor of up to 3.8. Evidence of a significant linkage to increased terrestrial plant productivity in the Northern Hemisphere is the increase in amplitude of the annual oscillation in the CO_2 level as recorded at the Mauna Loa station in Hawaii (Bacastow et al., 1985; Keeling et al., 1989). When the first records were made in 1959, there was a 6.5 ppm difference in CO_2 concentration between summer and winter. In 1992, the difference was 7.5 ppm. This suggests that during the major growing seasons during 1992, the plants of the Northern Hemisphere, where most are located, have sequestered up to 15% more carbon than they did in 1959. A close linkage is thus postulated between the observed amplitude increases and an ever increasing terrestrial plant activity (Keeling, 1983). It is further reported (Gates, 1985) that the seasonal amplitude of the atmospheric CO_2 concentration at various stations throughout the world varies with latitude, but all stations appear to have a slowly increasing amplitude with time. Whereas many factors could result in increases in the CO_2 amplitude at Mauna Lao, such as seasonal atmospheric circulation, afforestation and deforestation, seasonal fossil fuel use, and increases in ocean productivity, the most likely CO_2 seasonal behavior reflects an increase in global photosynthetic activity. *The overall increase of the amplitude of the annual cycle by about 10% between the decade of the 60s and the decade of the 80s gives direct evidence of an increase in the activity of the biosphere by about 0.5% per year.*

Rogers and Dahlman (1993) suggested a range of plant responses to CO_2 enrichment ranging from the biochemical to large-scale cropping regions. We have just addressed global impact response. The direct effects of elevated levels of atmospheric CO_2 are, however, most apparent on the overall individual plant performance and relative distribution, size, and quantity of various plant parts. They are visually evident and easily measured, especially under controlled environments. Additionally, they have constituted the majority of reports on the effects of atmospheric CO_2 enrichment. These responses (Table 4.2) have been observed, classified, recorded, summarized, and categorized by several investigators (Acock, 1990; Acock and Allen, 1985; Acock and Pasternak, 1986; Adams et al., 1990; Allen et al., 1990; Cure, 1985; Gifford, 1988; Goudriaan et al., 1990; Idso and Idso, 1994; Kimball, 1985; Lemon, 1983; Oechel and Strain, 1985; Rogers et al., 1983a, b; Rose, 1989; Sionit and Kramer, 1986; Strain, 1985, 1987; Strain and Cure, 1985; Waggoner, 1984; Wittwer, 1985, 1986; Woodward, 1987; Woodward and Bazzaz, 1988). Some detailed studies have been conducted on the distribution of plant parts with soybeans (Allen et al., 1991a; Rogers et al., 1992a, b), rice (Baker and Allen, 1993a, b; Baker et al., 1990, 1992a, b), and cotton (Kimball et al., 1993; Rogers et al., 1992b). Particular attention has recently been given to roots and the rhizospheres of cotton and soybeans (Rogers et al., 1986, 1992b).

Most experiments with economic crops involving the direct effects of elevated levels of atmospheric CO_2 have revealed two main overriding responses: (1) increased rates of photosynthesis, i.e., greater carbon fixation, and (2) an enhanced water use efficiency (Table 4.2). With these two vital processes, the impact of elevated levels of CO_2 has a positive effect on crop productivity. The one, increased water use efficiency, relates directly to an enhancement of food production through alleviation of a primary stress, and the other, an increase in photosynthesis, results in substantial boosts of yields and total agricultural output. Plants simply get bigger faster.

First, and foremost, is an increase in leaf and canopy photosynthetic rates, which increase with increases in CO_2 concentrations up to at least 1000 ppm, by volume, for many species. In general, the response follows the law of diminishing returns with little further response above about 1000 ppm. This range of effectiveness of CO_2 enrichment for enhancement of the growth of greenhouse grown lettuce, tomatoes, and cucumbers was observed over 30 years ago in Michigan (Wittwer and Robb, 1964). There are differences among species. For example, Yoshida (1976) concluded that for rice, the optimum CO_2 concentration for growth and yield was between 1500 and 2000 ppm. Further, photosynthesis of C_3 species has a qualitatively greater response to elevated levels of CO_2 than C_4 species. Significant increases in leaf areas occur with most plants with increases in weight per unit leaf area, leaf thickness, stem height, branching, root growth, seed, and fruit size and number. With both rice (Baker et al., 1990) and wheat (Gifford, 1977), there was a significant increase in tillering. Most plants respond by producing more leaf area per plant at faster rates at elevated levels of atmospheric CO_2. Rogers et al. (1983a, b) applied increments of 520, 718, and 910 ppm of atmospheric CO_2 to soybean (a C_3 plant) and maize (a C_4 plant) and noted the partitioning of dry matter to the vegetative

Table 4.2. Responses of Economic Crops to a Doubling of Ambient Levels of Atmospheric CO_2

I. Photosynthesis
   A. Rates are higher with few exceptions.
   B. Increases with higher temperatures and elevated CO_2.
   C. Light use efficiency is markedly increased with light deficiency.
   D. The light level at which photosynthesis balances respiration is reduced. The length of the photosynthetic day is thus increased.
   E. The response to lower light intensities increases photosynthesis for the understory of crops, forest, and other ecosystem canopies.
   F. Crop species with the C_3 pathway, show a greater response to elevated CO_2 than do plants with a C_4 pathway.
   G. Acclimatization, in which photosynthesis declines during long-time growth, most frequently occurs when rooting volumes are restricted or other environmental variables (low temperatures) progressively restrict growth.
   H. Differential responses of plant species suggest relative competitiveness will be altered.
   I. Increased photosynthetic CO_2 fixation has positive implications for food production, forest output, and storage in ecosystems.

II. Respiration
   A. Both photo- and dark-respiration are suppressed. Thus, both nocturnal and daylight enrichment with CO_2 are beneficial.

III. Water-Use Efficiency
   A. An increase in water use efficiency (carbon gained per unit of water lost) up to a factor of two.
   B. Both stomatal conductance, resulting from a narrowing of the stomates, and transpiration are reduced. Stomatal conductance of water vapor may be reduced by 40%. Both secondary effects benefit not only C_3 plants, but those of C_4 photosynthetic metabolism, which are mostly saturated at ambient CO_2.
   C. Reduction in transpiration is beneficial for both C_3 and C_4 plants.
   D. Water loss per unit leaf area and per unit land area is reduced.
   E. All plants benefit through improved water status.
   F. Stomatal density is decreased. Rice and a few other species are exceptions.
   G. Water use efficiency under field conditions may be considerably less than in more controlled environments.

Table 4.2 Continued

IV. Growth, Dry Matter Accumulation, Crop Yields, Total Output
   A. The impact on growth of most all plants is positive.
   B. For C_3 plants there is, on average, a 33% increase and, for C_4 plants, a 10% increase in economic yield or biomass production.
   C. Root growth is increased and most always very positive as to size or volume, architecture, micromorphology, and physiology. Root and tuber crops show dramatic yield increases.
   D. Root-soil microbes (bacteria including nitrogen fixers; and fungi, including mycorrihizae) are favorably influenced.
   E. Increased tillering in cereal grains and grasses.
   F. Faster growth and earlier maturity are usually important in all areas, but especially in regions where growing seasons are marginal in length for either single or multiple cropping. Earlier flowering in most instances.
   G. The sink for CO_2 and potential growth is greater for plants growing in open fields with greater rooting volumes to explore, and in a more turbulent atmosphere, than for plants confined in containers in a controlled environment. For widely spaced citrus trees biomass yields have been increased by over 200%.
   H. For the most extensive and longest (7 years) ecosystem study, so far (the brackish marsh of Chesapeake Bay), photosynthesis was stimulated, dark respiration and water loss reduced, and total carbon accumulation increased.
   I. Plants originating from high altitudes show a greater stimulation of growth by CO_2 than those from low altitudes.
   J. Both crop structure and physiology are markedly altered. Plants grow larger and accumulate more carbohydrates, and with legumes, more nitrogen as well.
   K. The harvest index is increased in most crops. A possible exception is soybean.

V. Environmental Stresses and Environmental Interactions:
   A. Temperature. The greatest increases in plant growth occur in environments having the highest temperatures. The least increases occur in the arctic. There is almost always, a positive effect on growth with a rise in temperature. More water is required per unit land area (response, however, is not universal). Flower and seed development may be either decreased, increased, hastened or retarded, depending on the normal species range and genetic flexibility of adaption.

Table 4.2 Continued

- B. Drought. A reduction in stress, related to an increase in water use efficiency for all plants, and superior root systems.
- C. Salinity. A reduction in stress, related to an increase in water use efficiency and improved rooting.
- D. Mineral stresses, excesses or deficiencies. There is usually a reduction at elevated levels of CO_2. Elevated levels of CO_2 will often partially compensate for a lack of soil nutrients. A positive plant response to CO_2 occurs over a wide range of nutrient availability. With increases of nutrient availability the CO_2 response appears to grow larger.
- E. Air pollutants. Harmful effects are partially alleviated with partial closing of stomates. The narrowing of stomates with increased levels of CO_2 infers possible protection from air pollutants that enter leaves by this route.
- F. UV-B radiation. There is a reduction in stress response.
- G. Competing biological systems—weeds, insects, pathogens. Marked alterations can be expected with the incidence of all competing systems. Interactions are not fully understood.

VI. Plant Composition

- A. Lower tissue and leaf nitrogen concentrations, with higher C/N ratios. May be significant for sugar crops. Exceptions are the legumes and other plants with symbiotic biological nitrogen fixation abilities.
- B. The nutrient components on surfaces of plant roots that attract and interact with microorganisms and other biota can be expected to change. There will be changes in plant microbe interactions, including biological nitrogen fixers and mycorrhizae. Most effects noted thus far are positive.

organs. There were significant percentage increases over ambient levels of CO_2 at 340 ppm in roots, stems, leaves, and total shoots of both plants. The most pronounced was in the soybean at 910 ppm of CO_2 and in maize at 520 ppm. The most significant gains were in the soybean roots with plants grown at 910 ppm CO_2.

The responses of roots and certain components of the rhizosphere to elevated levels of atmospheric CO_2 are particularly striking. Acock and Allen (1985), in a review of some 200 research papers on plant and crop responses of agricultural species to high CO_2 concentrations, stated that virtually nothing was then known about root growth response, other than there was usually an increase in root-to-shoot ratios. One of the few, and perhaps the earliest, report was that of Tognoni et al. (1967) using solution cultures of tomatoes and beans. Meanwhile, substantial increases in root dry weights, specifically for winter wheat (Chaudhuri et al., 1990), sorghum (Chaudhuri et al., 1986) and soybean (Del Castillo et al., 1989; Rogers et al., 1986), have been reported when these crops were grown under higher than ambient atmospheric CO_2. The increases in root growth were dramatic. Rogers et al., in two recent reports on two major agricultural crops, one with soybeans (1992a) and the other with cotton (1992b, c), found almost parallel results. The results of two levels of atmospheric CO_2, ambient 350 and 700 ppm, for soybean in a controlled (phytotron) environment were observed at Duke University. Cotton was grown under field conditions near Yazoo City in Mississippi, and near Maricopa, Arizona, under free air enrichment (FACE) at ambient 360 ppm atmospheric CO_2 and 550 ppm supplied during approximately 80% of daylight hours.

Responses with soybean from CO_2 enrichment, demonstrated substantionally positive effects on root system architecture, micromorphology, and physiology. Although there were substantial increases in the growth of stem and aerial plant parts, the effects on roots were even greater. Root lengths were increased by 110% and dry weights by 113%. Although root numbers did not increase, the diameter, stele diameter, cortex width, root/shoot, and root/weight ratios all increased. Results with CO_2-enriched cotton also showed significant increases in above ground growth parameters in height, diameter, and dry weights, both at Yazoo City, Mississippi, and Maricopa, Arizona, with even more striking effects than for soybeans on all root parameters (Prior et al., 1994a, b). Differences at the Maricopa site for roots showed dry weight increases for tap roots of 82% and for laterals, an attached length increase of 100%, and a 157% increase for dry weight. The findings clearly suggest that elevated atmospheric CO_2 stimulates root proliferation in both soybean and cotton, and also revealed that CO_2 enrichment for these two major crops enhances root growth more than that of shoots.

The one fact that is evident from the available information is that increasing levels of CO_2 in the earth's atmosphere will have virtually no adverse effects on plant root growth or function, and, indeed, roots will likely be positively affected in numerous ways (Table 4.2), which should benefit the health and productivity of most plant species (Krupa et al., 1993; Rogers et al., 1994a, b).

Increased rooting from CO_2 enrichment poses many interesting possibilities for both managed and natural ecosystems, including forestry (Mousseau, 1993). A 50% increase in root growth of pasture grasses has been reported (Newton et al., 1994), along with a significant increase in net photosynthesis and a proportional increase in roots for *Populus grandidentata* (Zak et al., 1993). Stimulation of rooting induced by CO_2 enrichment could be important, especially under moisture stress, in establishing crop and forest tree seedlings and transplants. Deeper soil penetration offers the probability of access to deeper reserves and the mining of more water as well as other soil resources. Greater root growth and penetration of soil profiles also suggest possibilities of improved plant mycorrhization and could translate into greater rhizodeposition (Rogers et al., 1992a, b; 1994). Clearly, future research must consider not only how CO_2 enrichment affects crop rooting, but also, how it impacts other, both physical and biological, processes (Prior et al., 1994).

The impact of elevated levels of atmospheric CO_2 on enhancement of yields of root and tuber crops appears to be somewhat selective for those staples that provide food for the earth's human population. Several of the major food crops (Irish potato, sweet potato, cassava, sugar beet, taro, yams) fall within this category and are among the 20 crops that stand between people and starvation (Wittwer, 1981a). These root and tuber crops are among those that show the most dramatic response of all crops in yield of harvested food products from CO_2 enrichment (Arteca et al., 1979; Baker and Enoch, 1983; Chapman and Loomis, 1953; Collins, 1976; Imai and Coleman, 1983; Imai et al., 1984). Further details of the responses of root and tuber crops to CO_2 enrichment will be given in Chapter 5.

It is now being reported that a common response of plants to an enrichment of CO_2 in the air is an increased root to shoot ratio. If such a shift from tops to roots is occurring worldwide, the result would be a most remarkable twofold—perhaps threefold—phenomenon resulting in a reduction of water required by plants, an increase in the carbon reservoir of the soil, and an overall increase in yields of major food crops especially the roots and tubers.

4.8 BIOLOGICAL NITROGEN FIXATION

An important direct affect of elevated atmospheric levels of CO_2, closely related to the enhancement of root growth and proliferation, is at the microbiological level and in the rhizosphere. CO_2 enrichment dramatically increases the amount of nitrogen fixed by leguminous crops. A doubling of CO_2 concentrations, producing more than a doubling of biological nitrogen fixed by the soybean, is not unusual (Hardy and Havelka, 1975, 1977). More recent studies (Williams et al., 1981) have confirmed the earlier reports.

This positive direct effect of CO_2 enrichment results in greater photosynthetic carbon fixation, and, for legumes, provides the energy that facilitates atmospheric nitrogen fixation by root symbionts. Legume/bacterial symbiosis is significantly increased by elevated CO_2 levels (Reardon et al., 1990). The increase is from more biomass (Acock, 1990; Brun and Cooper, 1967). Plants are larger and more carbon is allocated for nitrogen fixation. Since the growth of legumes, particularly root growth, is also greatly enhanced by CO_2 enrichment, this could enable such plants to expand their ranges with new habitats. A higher CO_2 world should not diminish the availability or cost of nitrogen fertilizers, but bring an improvement in both.

Many forest and range species, in addition to legumes, also have microbiological associations for facilitating biological nitrogen fixation, whereby such ecosystems would receive benefit. It has also been suggested that the mycorrhizae—fungal extensions of roots—common with many agricultural crops and forest tree species would be stimulated by the greater amounts of carbohydrates and other organic components exuded by roots of plants grown at higher atmospheric CO_2 levels. Such mycorrhization should improve the efficiency of nutrient uptake, particularly of phosphorus and some trace elements and soil moisture, thus increasing the efficiency of fertilizer use and the ability of plants to exploit soil resources of nutrients and water (Table 4.1) (Lamborg et al., 1983). Such capabilities have led Newman (1989) to declare that increased atmospheric CO_2 for farmers is an important aspect of the future. Growers and food producers can be expected to adjust their positions to take advantage of the CO_2 subsidy. Likewise, for forestry, it has been recommended that representatives of the forest industry begin developing plans now to adapt to a CO_2-altered world (Layser, 1980; Hoffman et al., 1982).

Concerning the fact that most plants respond favorably with growth increases from atmospheric CO_2 enhancement and there is an overwhelming positive effect on yield (Kimball, 1986b), most authors dealing with the direct effects of CO_2 having an agricultural orientation have concluded that the impact on crops will be positive in terms of vegetative extension, stem thickness, leaf area and size, seed and fruit production, and especially root growth and extension and the accumulation of dry matter. Yet, there are some ecologists quick to point out the few exceptions of little or no growth and an occasional negative response (Bazzaz and Fajer, 1992; Korner and Arnone, 1992; Oechel and Strain, 1985).

4.9 MUTUALLY COMPENSATING GROWTH FACTORS

A further point also seems invariably to be emphasized by some ecologists, in that such increases in productivity and yield of agricultural crops come about only under conditions where all other growth factors—water, light, temperature, mineral nutrients—are in ample supply, do not limit growth, and there is little, if any, competition from pests (weeds, insects, diseases). This is considered to be true for crops grown both in controlled environmental facilities and in the field. The implication is that favorable responses to atmospheric CO_2 enrichment occur only in an agroecology setting, and not in natural ecosystems where plants have to compete with each other, not only for CO_2 but water, space, soil nutrients, and sunlight. The frequent implication is that in greenhouses and field crop agriculture all other factors for plant growth are at optimal, or near optimal levels and are seldom limiting. Such is pure speculation (Morison, 1993). This erroneous concept of limiting factors was rigidly held by many workers and was given even further support by Kramer (1981), whose paper is widely acclaimed by some ecologists and climatologists. The adverse effects of CO_2 enrichment reported were based on poorly designed, conducted, and interpreted experiments of other researchers. Yet the review by Kramer led a large audience to believe a downgrading and loss of photosynthetic capacity followed CO_2 enrichment. This paper is still being referenced by a leading and vocal ecologist (Woodwell, 1993).

Kimball (1986b) aptly stated that any discussion of environmental constraints almost inevitably leads to a discussion of the concept of the Law of the Minimum or the Principle of Limiting factors. This was first proclaimed by Blackman (1905) and has been incorporated into many texts of plant physiology. Simply stated, the principle is "when a process is conditional as to its rapidity by a number of separate factors, the rate of the process is limited by the pace of the 'slowest' factor." If, for example, a factor that is limiting growth is supplied at a faster rate, then growth would increase until another factor became limiting. Meanwhile, no further increase in growth would then be expected from supplying more of the first factor. The process would then be saturated when additional increments of supply produce no further growth. Although this principle may have some validity, plant growth is not that simple. In particular, this is the case in plant response to CO_2 enrichment. Among the very first to point this out were Hopen and Ries (1962), who emphasized the use of the term "Mutually Compensating Effects" rather than the Law of the Minimum for variable CO_2

concentrations and light intensities in the greenhouse culture of the cucumber. This report was preceded by a paper from Gaastra (1959) emphasizing the interdependence of CO_2, light, temperature, and stomatal resistance on the growth of several crops. Carbon dioxide at high atmospheric levels compensated for both light and temperature deficiencies. Further evidence of the "Mutually Compensating Effects" of elevated levels of atmospheric CO_2 and light and other growth factors, including temperatures, soil nutrients, and water, was summarized by Wittwer and Robb (1964) and incorporated into recent reviews by Wittwer (1985, 1990, 1992, 1994) and Baker and Allen (1994).

Downton et al. (1980) speculated, based upon their own studies of CO_2 enrichment on both C_3 and C_4 plants in Australia and those of Wong (1980) for cotton and corn, that the simultaneous increases in both photosynthetic rates and decreases in transpiration rates would counter, to some extent, the predicted increases in temperature and reduced rainfall for many important agricultural areas. Specifically, the temperature optimum for photosynthesis would be shifted upward. This, coupled with improved water use efficiency under CO_2 enrichment, may enable plants to occupy habitats that in the absence of such enrichment would be too arid and too hot to support life. For wheat plants grown under dry conditions, elevated levels of CO_2 in the atmosphere can be a partial substitute for water (Dyson, 1992).

Additionally, the "atmosphere" of the nature of plant responses to elevated levels of atmospheric CO_2 has been clouded by the ever changing positions of widely quoted and noted ecologists.

Strain (1985), in presenting a background on the response of vegetation to atmosphere CO_2 enrichment, stated:

> In addition, the questionable concept of single limiting factors continues to cloud the thinking of some researchers. It can still be heard in meetings and read in literature that CO_2 enhancement will not have an effect on plant growth because some other factor is limiting. No environmental factor operates alone. All factors in the actual environment (Billings, 1970) of an organism are potentially active. A change in any factor will modify the physiological behavior of an organism. The degree and nature of the modification varies with developmental stage and with genetic variability. When the ambient CO_2 level changes around a plant, the physiological response to other factors will be modified. This concept of holocoenosis (Billings, 1970) is beginning to be more generally accepted by biologists, but the erroneous idea of single limiting factors still prevails.

It is surprising, however, to note in a recent editorial by Strain (1992) almost a complete contradiction of this concept in his declaration, speaking of the thousands of positive results in yield increases and crop productivity in experiments and experiences with elevated levels of atmospheric CO_2 that "all these observations were made under controlled agricultural conditions in which all other environmental requirements were maintained at optimal levels. In other words, when temperature, light, soil water, and mineral fertilizers are present in optimal ranges and when plants are not suffering from disease, weed competition, or herbivory, adding carbon dioxide to the atmosphere generally increases plant growth and crop yield." Further, it seems his latest conclusion is that increased levels of atmospheric CO_2 are beneficial only if other potentially limiting resources or environmental factors remain available and are optimal for the new CO_2 level.

This position is a complete departure from the results of hundreds of earlier studies and his own and other very recent reports. Virtually all authors that have considered or recognized the direct effects of CO_2 have concluded that the impact on crops will be positive, and that CO_2 can help ameliorate the environmental stresses or limitations of water, temperature, light, nutrients, salinity, and air pollutants (Idso and Idso, 1994). All these constraints have been observed to interact with elevated levels of atmospheric CO_2 (Table 4.2).

To repeat, environmental conditions are seldom ideal, even in a greenhouse or other controlled environment facilities. Light is always limiting when the leaf area index (LAI) gets high enough. Kimball (1985) summarizes results from many experiments by stating it appears that at low light intensity, decreases in the light compensation point with increasing CO_2 concentration results in huge relative increases in photosynthesis of C_3 plants. Also large relative and absolute increases could also occur at high light intensities because of saturation effects. Idso (1989a, b, 1990), with supporting evidence from studies other than his own, has suggested that surprising as it may seem, when the plant environment is not conducive to optimum plant growth, with limitations of light, water, temperature, or nutrients, the relative stimulating effects of atmospheric CO_2 enrichment are even greater than they are under optimum conditions.

With agriculture in open fields, growth conditions are seldom, if ever, ideal for any extended period of time. Even a single crop is usually exposed during its growth cycle of one season, to both shortages and surpluses of soil moisture, adverse high or low temperatures, a lack of sunlight, severe storms, and invasions of weeds and pests.

The compensating effects of elevated levels of atmospheric CO_2 for light, temperature, water, and mineral nutrients have been summarized in several recent reports (Bunce, 1992; Idso, 1989b; Idso and Idso,

1994; Kimball, 1985; Morison, 1993; Rogers and Dahlman, 1993; Wittwer, 1990), and stand as a body of evidence and witness against the single factor plant growth approach. The results of continuous CO_2 enrichment on field grown cotton, at close to natural conditions in a free-air field environment, where both optimal and limiting levels of water and nitrogen were maintained, are very revealing (Kimball et al., 1993b). The data show that for five continuous growing seasons and in all experimental treatments, including limitations of water and nitrogen, cotton yields and dry plant weights were increased significantly by CO_2 enrichment.

Johnson et al. (1993) and Rozema (1993) have perceptively emphasized that the changing CO_2 levels on vegetation of the past, present, and future, all of which are now in process, should be viewed as a powerful influence, operating sometimes simultaneously and sometimes independently of shifting temperature, light, and precipitation patterns or regimes. Idso and Idso (1994) and Idso et al. (1986, 1987) concluded from the response of many species that plant growth from increased CO_2 was greater at higher temperatures. The effects of changes in temperature on plants, as predicted for global warming, will be different with those originating in hot arid regimes from those in cold arctic tundra.

In summary, when atmospheric CO_2 concentrations are increased, both the photosynthetic CO_2 uptake rates and the temperature optimum for maximum photosynthesis increase (Allen and Boote, 1992). Moreover, the theoretical analyses based on an understanding of the interaction of CO_2 and temperature at the level of carboxylation suggest that photosynthetic CO_2 uptake will be stimulated by rising levels of atmospheric CO_2, especially in warm climates (Table 4.1). This will, in most instances, be amplified by any simultaneous increase in mean temperature of the order predicted by current climate scenarios.

With the current projections for a global warming, that presumably will accompany rising levels of CO_2, it may be good fortune that there is a strong interaction between plant response to increased temperature and CO_2 concentrations. Long (1991) calculated that the optimum for C_3 plants, which comprise about 95% of the earth's plants, may be raised by 5°C for crops at elevated CO_2 levels (350 vs. 650 ppm). His findings question the value of models, which are being widely used, for predicting plant production in response to projected climate change, which ignore the direct effects of rising CO_2, and the modifications that rising CO_2 imposes on the temperature response of net CO_2 exchange. Idso and Idso (1994), after a thorough review of all experiments thus far conducted, conclude that the relative growth-increasing effects of atmospheric CO_2 enrichment typically rise with increasing air temperature over the entire temperature ranges investigated. The optimum temperature for plant growth generally rises with atmospheric CO_2 enrichment.

A global assessment of the effects of CO_2 increase on agroecosystems and vegetation requires careful analysis of the interaction between CO_2 enrichment and temperature. A CO_2-induced climate change comprises the simultaneous increase in atmospheric CO_2 and perhaps global warming. Experimental data are alarmingly scarce on these joint effects (Kimball et al., 1993a, b). Not only will photosynthesis and respiration be affected by changes in temperature and CO_2 levels, but many other metabolic processes. C_3 plants may be more favored by atmospheric CO_2 enrichments, and C_4 plants more by global warming. As reported, CO_2 enrichment increases the growth of many plants under low light, but C_3 and C_4 plants could be expected to differ in their responses. Photorespiration is suppressed in C_3 but not C_4 species, and, in general, reduced growth rates from water stress can be counteracted by CO_2 enrichment (Rozema, 1993; Rozema et al., 1993).

4.10 GREENHOUSE CROP PRODUCTION

The global production of high value food and ornamental crops in greenhouses and other protective structures has grown enormously in area, and advanced greatly in technology during the last quarter of the 20th century (Wittwer, 1993; Wittwer and Castilla, 1995). The use of plastic as a covering and as an alternative to glass was first introduced in America in the early 1950s (Emmert, 1955). Plastic also has an advantage over glass for CO_2 enrichment of greenhouse atmospheres, because it more effectively retains added CO_2.

The beneficial effects of elevated levels of atmospheric CO_2 were first noted in the culture of greenhouse crops in northern Europe during the latter part of the 19th century (Van Berkel, 1986; Wittwer and Robb, 1964). It was an experiment, however, conducted by Van Helmont (1648) that set the stage for the role of CO_2 in crop production and contributed to the discovery of photosynthesis. He planted a 5-lb willow tree in a bucket of 200 lb of oven-dried soil. Water was added as needed. After 5 years, the tree had increased to 169 lb, 3 oz., but the soil itself still weighed 199 lb and 14 oz. This experiment should forcibly remind us that the primary building material from which the bulk of a plant originates is the CO_2 in the atmosphere.

Today, there are approximately 275,000 ha of high value horticultural crops grown in plastic greenhouses or high tunnels and a lesser number, perhaps 41,000 ha of glass or fiberglass greenhouses (Wittwer and Castilla, 1995). Enclosed areas have more than doubled during the decade of the 1980s. The most phenomenal growth has been in the Mediterranean countries of Spain, Italy, Greece, Turkey, Egypt, and Morroco, and in the oriental countries of China, Tawain, Korea, and Japan. There is the potential in each greenhouse for CO_2 enrichment (Table 2.2).

Extensive experimentation with enrichment of greenhouse atmospheres for the production of vegetables (tomato, pepper, eggplant, cucumber, lettuce) and flowers (carnations, roses, chrysanthemums, poinsettia, snapdragons, bedding plants) was conducted in the late 1950s, during the 1960s, and in the early 1970s. The results of these studies have been summarized by Bauerle et al. (1986), Enoch and Kimball (1986), Goldsberry (1986), Hanan (1986), Moe and Mortensen (1987), Porter and Grodzinski (1985), Slack (1986), and Wittwer (1986). With vegetable crops, the effects are more rapid growth, earlier maturity, larger fruit size, greater height and total yields, and an improvement in fruit quality. With flowers, there is earlier flowering, an increase in the number of fancy grade flowers, increases in flower stem lengths, overall yield increases, and a reduction in time to grow a marketable crop. The general procedure in CO_2 enrichment for greenhouse crops is to maintain two to three times the ambient level of atmospheric CO_2 from about sunrise to 1 to 2 h prior to sunset. Atmospheres, generally, are not enriched during the night. A good target choice for an optimum concentration, as suggested by our initial studies (Wittwer and Robb, 1964) and those of others as summarized by Kimball (1986a), is 1000 ppm. Thus, the projected CO_2 concentration of the global atmosphere of the future resembles that being deliberately created by today's greenhouse crop growers (Kimball et al., 1993a).

Whereas the benefits of CO_2 enrichment are well known among commercial growers of flowers and vegetables in enclosed greenhouses, which are often depleted in CO_2, resulting in levels considerably below the ambient atmosphere in the open, CO_2 enrichment is not universally practiced for increasing commercial crop production.

The reasons are several. Most modern greenhouse crops are grown and marketed during the spring and early summer, when it is neither desirable nor feasible to keep ventilators closed to retain the released CO_2 during the daylight hours. The equipment to produce, release, distribute, and monitor the released CO_2 represents a significant investment. Finally, with adequate ventilation ambient or near ambient levels of CO_2, which continue to increase, can be maintained within most greenhouses. Repeated experiments have, however, demonstrated, beyond doubt, the value of CO_2 enrichment in the production of greenhouse crops and this value has been vindicated by current acceptance and practice by many successful growers. The benefits include the multiple effects of increased vegetative growth, higher yields of improved quality, and hastening of maturity. There are also compensating effects relating to deficiencies of light, temperature, soil moisture, and mineral nutrients. Optimal levels of CO_2 enrichment for most greenhouse crops range from 800 to 1200 ppm. The most important food crops currently grown in greenhouses, all of which respond to elevated levels of atmospheric CO_2, are cucumbers, tomatoes, strawberries, peppers, eggplant, green beans, melons, muskmelons and watermelons, squash, Chinese cabbage, and lettuce. The most important flower and ornamental crops are bedding plants, roses, geraniums, chrysanthemums, carnations, and gerbera. Down-regulation, also known as acclination, of the photosynthetic capacity, as reflected by decreased yields induced by continuous exposure for extended periods of time to elevated levels of atmospheric CO_2, has not been noted.

Experiments are still being conducted on CO_2 enrichment of greenhouse grown crops and reports are being published (Hicklenton, 1988; Lorenzo et al., 1990; Mortensen, 1987; Nederhoff, 1990, 1994; Nederhoff and Vegter, 1994; Nederhoff et al., 1991; Van Huylenbroeck and Debergh, 1993; Vermeulen and van de Beck, 1993; Willits and Peet, 1989; Wittwer, 1992). I stated over two decades ago (Wittwer, 1970) that sufficient information had been accumulated at that time on the benefits of CO_2 enrichment on greenhouse-grown crops, that further experiments were unnecessary. Papers summarizing the effects of CO_2 fertilization on crop plants and the complementary or compensating effects on concomitant limitations of light, temperature, moisture, and nutrient deficiencies have since been published periodically (Wittwer, 1978b, 1985, 1990, 1992, 1994; Wittwer and Homma, 1979). The conclusions of the 1970 report continue to be supported by subsequent research and grower practices.

Today, in Holland, all crops, vegetable and ornamentals, grown in all the glass greenhouses covering over 10,000 ha and with an annual return of over $3 billion, are enriched with atmospheric CO_2 at a level approaching 1000 ppm during daylight hours, and during the entire year when crops are produced. Increases related to enriched atmospheric levels of CO_2 for marketable yields of tomato, cucumbers, sweet pepper, eggplant, and ornamentals range between 20 and 40% (Nederhoff, 1994).

5

Global and Regional Projections of a CO_2-Induced Climate Change and Food Production

5.1 INTRODUCTION: FOOD PRODUCTION AND THE RESOURCE BASE

Globally, some 25 crops stand between people and starvation (Wittwer, 1981a). The production of these crops, their climatic adaptability, and their direct response to CO_2 holds priority in any assessment of the effects of currently rising levels of atmospheric CO_2 on agricultural output, world food production, and global food security (Table 5.1).

The largest single food group is the cereal grains (rice, wheat, corn, barley, oats, rye, sorghum, and millet), which provide approximately 60% of the calories and 50% of the protein consumed by the human race. Three of them, in order of importance—rice, wheat, and corn, supply almost a third of the world's food, and approximately 85% of the world's cereal exports. They are the staffs of life, along with potatoes and soybeans (Kahn, 1985). The legumes (soybeans, field beans, cowpeas, chickpeas, pigeonpeas, mungbeans, broadbeans, urd beans, peanuts, garden peas, lentils) provide about 20% of the world's protein for human consumption. The soybean, alone, accounts for about 67% of world trade in protein cake equivalent. Food animals provide 20% of the protein for people, with 5% coming from fish. The balance of calories, protein, and essential vitamins and minerals is derived from the tuber (Irish potato) and root (cassava, sweet potato, yams, cocoyams) crops and three tropical crops, coconuts, bananas, and plantains, followed by fruits, vegetables, and nuts, including pineapple, citrus, grapes, apples, tomatoes, melons, and cabbage. There are over 100 food crops produced in the United States, each with an annual farm value of

TABLE 5.1. Geographic Distribution of Food Crops and Projected Changes in Yields Resulting from a Doubling of Ambient Levels of Atmospheric CO_2

Crop	Current latitudinal and temperature distribution	Projected changes in distribution and yield: Little or no temperature increase	Modest temperature increase
Rice	Latitude 53°N to 35°S throughout tropics	No change except higher yeilds	Both N and S boundaries extended; seed yields may be reduced in tropics
Wheat	Temperatures ranging from –40°C to +40°C; not in lowland humid tropics	Drier, less humid areas, higher yields	Both N and S boundaries extended; production restrained in tropics
Corn	55°N to 45°S, from sea level to 4000 m and throughout	Boundaries extended into drier areas; same yields or higher	Both N and S boundaries extended; no loss in tropics
Grain, sorghum, pearl, millet	55°N to 45°S and throughout the tropics; will grow at wider range than corn	Boundaries extended into drier areas beyond corn; same yields or higher	Both N and S boundaries extended; no loss in tropics
Barley, oats, and rye	In temperate zones to 60°N to 60°S; Barley is earliest in maturing and most adaptable	Boundaries extended; higher yields, earlier maturity	Production zones move both further north and south
Potato	In almost every inhabited country, state and province in the world at some season or elevation	Boundaries extended farther with increased yields	Boundaries extended even farther, except low elevations in the tropics
Sweet potato	40°N to 35°S throughout humid tropics in rain-fed areas; temperatures above 10°C	No boundary change, but much higher yields and earlier maturity; more edible root	Both N and S boundaries extended with no change in the tropics; more edible root
Cassava, yams, cocoyams	30°N to 30°S mean temperature greater than 20°C and 750 ml rainfall; no areas too hot	Little change in distribution but significant increases in yield; more edible root growth	Boundaries extended north and south; yields increased; more edible root growth

Table 5.1 (Continued)

Soybean	Widely distributed 55°N to 45°S and throughout the tropics, chiefly between 25 and 45°C and altitudes of less than 1000 m	No change in distribution but significant yield increases; increased biological N_2 fixation; more root growth	Boundaries slightly extended; small increase in yields; increased biological N_2 fixation
Field beans	Widely distributed, 50°N to 45°S; elevations throughout tropics at mean temperature of 19–23°C during flowering	Little change in distribution; yields increased; increased N_2 fixation	Boundaries slightly extended; no change in yields if temperatures are higher
Cowpeas	35°N to 30°S hot-dry and humid tropics—drought resistant	No change in distribution; yields and biological N_2 fixation increased	Boundaries slightly extended N and S; effect on yields neutral, or slightly positive
Chickpeas, pigeon peas, mung beans	Between 15 and 35°N latitude, mostly 20 to 30°N	No change in distribution; yields and biological N_2 fixation increased	No change in distribution; yields and biological N_2 fixation increased
Peanuts	38°N to 35°S and in tropics; required heat sums 2500 to 4800°C	No change in distribution; yields increased	Boundaries extended both north and south
Rapeseed (canola), sunflower	Widely distributed, mostly from 25° to 60° latitude north and south	No change in distribution; yields increased; projected yield increase	Boundaries extended both north and south
Plantains, bananas, coconuts	In lowland and highland tropics; temperature tolerance not below 10–12°C; 30°C and 100 mm rainfall/month the best; no place too hot	No changes; no information on yield	Boundaries extended north and south; no information on yield
Sugarcane	36.7°N to 31°S from below sea level to 1000 m; no place on earth too hot; 26–30°C best	No change in distribution; yields increased	Both north and south boundaries extended; effects on yield not known

TABLE 5.1. Geographic Distribution of Food Crops and Projected Changes in Yields Resulting from a Doubling of Ambient Levels of Atmospheric CO_2 (Continued)

		Projected changes in distribution and yield	
Crop	Current latitudinal and temperature distribution	Little or no temperature increase	Modest temperature increase
Sugar beet	38 to 52°N latitude; North America, Europe, China, India	No change in distribution; yields increased	Boundary extended northward; effects on yield not known
Deciduous fruits, nuts, small fruits, grapes	30 to 55°N, 30 to 45°S latitude, temperate zones; high elevations in tropics	No change in distribution; yields increased	Boundaries extended toward poles, suppressed toward equator; effects on yield not known
Citrus and tropical fruit	Tropics and semitropics to 40°N and 30°S latitudes	No change in distribution; yields increased	Little change except N in Mediterranean area; effect on yield not known
Hardy vegetables	All temperate zones to 62°N and 45°S latitude and high elevations in tropics	No change in distribution; yields increased	Northern and southern boundaries extended; yields increased or no effect
Tender vegetables	Frost-free periods of 60–120 days; temperate and subtropical, 50°N to 40°S latitude	Boundaries slightly extended; yields increased	Northern and southern boundaries extended; yields increased, decreased, or no effect

over a million dollars. In China, there are over 125 species of vegetables alone, grown for human food.

The major inputs for food production are climate, land, water, genetics, chemicals (fertilizers, pesticides), and technology (Crosson and Anderson, 1992). The cropped area of the world is now about 1.5 billion ha. There is the potential for much more, particularly in the United States, Russia, Brazil, Argentina, Central America, Australia, New Zealand, many African nations, and even in China, India, and Indonesia (Avery, 1991). There are more than 3 billion ha in pasture and rangeland, almost 4 billion in forest, and more than 4 billion in urban, and other nonagricultural or nonproductive uses (World Resources Institutes, 1992–93).

A recent report by Waggoner (1994a) emphasizes that advances in farming technologies, combined with changes in diets in response to health and price, will ensure that the world's population will use existing cropland and water more economically and efficiently and save more land for natural or wilderness use in the next 50 years, even with a population increase to 10 billion people. The global total of sun on land, CO_2 in the air, and fertilizer and water could produce far more food than 10 billion people need. Waggoner (1994a) goes on to affirm *"the silver lining in the growing cloud of atmospheric CO_2 that may warm the planet is more raw material for photosynthesis."* The CO_2 in the air is rising at the rate of 1 to 2 ppm/year, which will raise the atmospheric concentration by approximately one third while the population of the earth grows to 10 billion. There will be an increase in food crop production of about 10%.

Much attention is now focused on the climatic resource, probably the most determinant of all natural resources on agricultural productivity, because of its unpredictability and variability, and our inability to predict accurately, by computer modeling or otherwise, either temperature or precipitation patterns, both of which impact on global and regional food production. The inaccuracy of computer modeling projections of CO_2-induced temperature changes on a global scale and the still less certain regional changes in precipitation have added additional complexities to climate resources. It is no small source of amazement that the climate resource, the most determinant of all natural resources for agricultural productivity, is seldom considered as a variable, but a constant in projections of the global food supply. Many environmental and human health issues have arisen with respect to the agricultural use of fertilizers and pesticides. Overall, the world's food supply for the future may depend less on natural resources than on human knowledge, initiatives, and new technologies (Avery, 1991). The resource base changes with time and technology. There are many examples of the resource base being enhanced by new technology. This is true of land, water, energy, climate, atmospheric CO_2, and genetics.

Almost all studies of elevated levels of the earth's atmospheric CO_2, as they are projected for impacting future agricultural and food production, have dealt with varying computer-derived scenarios or historical analogs of increments of increasingly higher temperatures and shifts in precipitation patterns. These scenarios and analogs (Bach, 1979; Blasing and Solomon, 1982; Decker et al., 1985; Newman, 1982; Parry et al., 1988a, b; Hillel and Rosenzweig, 1989) have not incorporated in their modeling projections the adjustments the farmers would naturally make as new realities are comprehended. These included potential and inevitable improvements in agricultural technology and management, changes in the resource base, or, in many instances, the direct beneficial effects of CO_2 enrichment of making plants grow faster and tempering the transpiration of water (Patterson and Flint, 1990; Waggoner, 1990). Studies abound, consisting of, on the one hand, either the direct beneficial effects of CO_2 enhancement on crop yields and growth, ignoring any climate change, or, on the other hand, of projected climate change impacts on agricultural productivity, derived from the output of various general circulation computer models, resulting from a doubling or other elevated levels of atmospheric CO_2. For the real world of the future, the direct effects of constantly elevating levels of atmospheric CO_2 will, without doubt, prevail, alongside any induced climate change.

In almost all instances, when the direct effects of elevated levels of atmospheric CO_2 are combined with projected climate changes on plants and crops, the temperatures selected from most models have been higher than what has been experienced and recorded thus far in the real world. This is true for one of the latest global assessments (Rosenzweig et al., 1993), involving four of the major food crops of the world (rice, wheat, corn, soybean) in 18 countries. Namely, there have been significant increases in atmospheric CO_2 and other greenhouse gases with no definite climatic warming change (Balling, 1992; Idso, 1989a; Lindzen, 1993; Michaels, 1992, 1993). Moreover, the impact of the direct, mostly positive effects of the rising atmospheric CO_2 on crop production has either been ignored or minimized in innumerable reports on projected catastrophic climate change. Nevertheless, where the positive effects of CO_2 have been factored into climate change, they have offset some, if not all, of the unlikely projected deleterious effects of computer model projected climate change (Adams et al., 1990; Rosenberg, 1992a, b; Rosenzweig and Parry, 1994; Rosenzweig et al., 1993; Waggoner, 1992).

What climatic scenario of warming is used determines the predictive effects of factoring in the direct effects of the CO_2 on plant growth, crop yields, and increased water use efficiency. Where the Goddard Institute of Space Studies (GISS) general circulation model (GCM) was employed as a reference, field crop output was generally increased. Under the more severe Geophysical Fluid Dynamics Laboratory (GFDL), model field crop output was more often decreased with a doubling of atmospheric CO_2 (Rosenzweig, 1989). Similarly, Adams et al. (1990) used a general equilibrium economics model linked with atmospheric and plant science models for U.S. agriculture. They found the GISS scenario resulted in an 18% decline in crop prices and a $9.9 billion increase in economic surplus. With the GFDL-climate scenario, there was a price increase of 28% and the economic surplus declined by $10.5 billion. As a major policy item, it was concluded by Adams et al. (1990), future climate change for agriculture induced by rising levels of atmospheric CO_2 would not result in a food security problem in the United States.

There are now many reports covering the projected effects of the rising levels of CO_2, other trace gases, and climate change (warmer temperatures and shifts in precipitation patterns) on local, regional, national, and global agriculture. Some specifically address the situation in the United States and regions within the United States (Adams et al., 1989, 1990; Decker et al., 1985; Dudek, 1989; National Academy of Science, 1992; Rosenberg, 1992a; Rosenzweig and Hillel, 1993a; Smith and Tirpak, 1989a, b, c; Waggoner, 1983, 1990, 1992; Wittwer, 1980). Other reports have a global perspective (Hillel and Rosenzweig, 1989; Kane et al., 1992; Kimball et al., 1990a, b; Parry, 1990; Parry et al., 1988a, b; Rosenzweig and Hillel, 1993b; Rosenzweig and Parry, 1994; Rosenzweig et al., 1993). It has been concluded (Kane et al., 1992) that the effect of moderate climate change on world and domestic food economies may be small, as reduced production in some areas is balanced by gains in others.

The International Rice Research Institute (IRRI) and American Association for the Advancement of Science (AAAS), (1989), Kenny et al. (1993), and Shu Geng and Cady, (1991) address specific countries or regions within countries (Downing, 1992; MacCracken et al., 1990), or agriculturally developing or developed regions (Downing, 1992). Many of these voluminous reports, especially for the agricultural developing world, do not factor in the direct effects of elevated levels of atmospheric CO_2, or do it inadequately, on plant growth, food production, and improved water use efficiency (Downing, 1992; Kenny et al., 1993; IRRI and AAA, 1989; Parry, 1990; Parry et al., 1988a, b; Shu Geng and Cady, 1991). All reports are still speculative, seriously limited by the number of samples, and admit there are large differences between all models used.

Projected results of a global warming on food crop production were always highly dependent on the GCM or scenario used, and the extent of the projected temperature increase. For composite scenarios in Europe (Kenny et al., 1993), the only major land area without extensive deserts, the effect of temperature change, without considering the direct effects of CO_2, results were variable depending on the scenario, site location, and whether the crop was water limited. Increases of 2 or 4°C generally resulted in higher potential yields in northern Europe with no change in central Europe up to the year 2050. Potential yield decreases occurred in southern Europe because of water-limited production. Yield increases in northern Europe were greater for wheat than barley, and forage production was increased in eastern and western France under all scenarios. For maize yields, sensitivity analysis showed that potential grain production would increase in northern Europe, remain the same in central Europe, and decrease in southern Europe. An increase in temperature would increase the total area in Europe that would be water limited for crop production. For winter-sown carrots, an enrichment to 560 ppm CO_2 would increase root yields by 50% over a wide range of temperatures with no temperature–CO_2 interaction on yield. For Europe, all climate change scenarios showed a northward expansion into higher latitudes. There would be a progressive northward and eastward shift for viticulture, and it would become commercially viable in the United Kingdom (Kenny et al., 1993).

Hopefully, the following discussions will provide some insight as to the agricultural potential, during the 21st century, that would result from higher levels of atmospheric CO_2 and other trace gases on both the growth and yield of crops and on climate change. The importance of always incorporating the direct effects of CO_2 on crops to understand fully the effectiveness of adjustments in responding to greenhouse warming is underscored. Equally important, and even more unpredictable, will be the adjustments and adaptations that farmers will make as new realities appear and as improvements in technology and management become available. We will now consider the predictive responses and adaptiveness of the major food crops, the food animals, forests, rangelands, and aquatic systems to both climate and higher atmospheric levels of CO_2.

5.2 THE MAJOR FOOD CROPS

The Cereal Grains

Rice, wheat, and corn, viewed together, contribute about one-third of the global total food supply for the people of the earth. Rice and wheat are rivals as the number one staple food crop.

Rice

Rice is unique as a cereal grain on at least three counts. Unlike other cereal grains, it is used almost exclusively for human food. Only the byproducts go for animal feed or industrial uses. It is the only one that is cooked and eaten as a whole, usually polished, grain, and as a semiaquatic cereal, and the only one commonly grown in standing water. Most of the world's rice is paddy rice, which requires irrigation and a high degree of water management. Its water requirement is the highest among all the cereal grains. Rice is more sensitive to climate aberrations than other cereals because it is primarily a monsoon crop in the developing world with specific water requirements. About 90% of the world's rice crop is produced and consumed in monsoon Asia. This area extends across southeastern Asia from India to Japan and includes all the adjacent tropical and subtropical islands.

China, with 22% of the world's population, annually produces about 37% of the world's total of about 470 million metric tons. Over two thirds of the Chinese people depend on rice for their staple food. India ranks number one in area of rice culture, but second to China in the amount produced. Rice is grown in every province, municipality, and autonomous region in China; every state in India; throughout Indonesia, Thailand, Sri Lanka, the Philippines, Vietnam, Korea, Taiwan, Hong Kong, Japan, Bangladesh, Burma, Nepal, Singapore, and Malaysia; and in Brazil, Colombia, Dominican Republic, Guyana, Panama, Gambia, Guinea, Ivory Coast, Liberia, Madagascar, Mauritania, Sierra Leone, Nigeria; the Middle East; Southern Europe; and Australia. For the United States, it is an important crop in Arkansas, Mississippi, Louisiana, Texas, and California. It is the most important food crop for the agriculturally developing world and the main staple in the diets of the vast populations of south and southeast Asia, where 91% of the rice is grown. Rice is the basic food for more than half the world's population. It supplies more dietary energy than any other single food. In tropical regions, it is planted and harvested the year-round. It is one of the oldest cultivated crops on earth, and probably the world's most versatile crop. Over millennia, in different parts of the world, rice has been adapted to very different climatic areas.

Rice flourishes from the 53°N latitude along the northeastern borders between China, and Russia, to 40°S in central Argentina, and 35°S in New South Wales of Australia. It is one of the world's most widely adapted crops. It is cultivated in the cool climate at altitudes above 2000 m in Nepal, Kashmir, and China and in the hot arid climates of southern Pakistan and Iran. Rice is found growing below sea level in Kerala, India, and the Xinjiang autonomous region of China, and at sea level in Bangladesh and many other rice-growing areas. In Latin America, Africa, and parts of Asia, it is grown in upland areas frequented by drought. At another extreme, floating rice grows in water 3 m deep in Thailand, Bangladesh, Burma, and Vietnam (Seshu et al., 1989).

Rice, however, is notoriously temperature sensitive. Currently, its production is restricted by temperatures both too hot and too cold. Rice maturity is correlated with heat units or temperature sums related to variety, latitude, planting season, and photoperiod. With significant warming in the tropics, now, however, predicted for the hotter regions by GCMs, rice farming might have to shift to other crops (Walsh, 1991). Wetland rice fields may themselves make a significant contribution to global warming, as one of the major biological sources for the emission of methane (Heinz-Ulrick Neue, 1993). As noted earlier, methane, N_2O, CFC-11, and CFC-12 are making significant contributions to radiative forcing and are about equal in total to that of CO_2 (Wang et al., 1991b). A recent and careful modeling study of the effects of climate change on world rice production (Penning de Vries, 1992) reports insignificant changes in the warmest regions and possibly small, at most a few, percent gain in yields, but reports uncertain changes in other regions. The temperature response of rice varies with the stage of growth, variety, duration of the temperature, diurnal changes, and plant status (Yoshida, 1981; Yoshida and Pardo, 1976). Sterility of spikelets is caused by high temperature, almost exclusively on the day of anthesis (Satake and Yoshida, 1978). When temperatures exceed 35°C, the rice crop is injured. The degree of injury, however, depends on the type of rice, variety, and the stage of plant development. Elongation of internodes in deep water rice is retarded at 35 to 40°C. In fact, temperatures greater than 35°C for more than 1 h may induce a high percentage of sterility (Yoshida, 1981). Baker et al. (1992a, b, c) concluded that biomass, tillering, and grain yield in rice maintained in sunlit chambers and in paddy water may be increased by 60% or more, with daytime temperatures of 28°C, nighttime temperatures of 21°C, with paddy water at 25°C, and with atmospheric levels of CO_2 increased from 330 to 660 ppm. They emphasize, however, that the beneficial effect of CO_2 on rice productivity could possibly be negated if temperatures also rise and if this were to occur under low solar irradiance. Low solar radiation results in low productivity. Solar radiation regionally is the primary climatic determinant of yield potential in rice. In the tropics, dry season rice yields usually exceed that of the wet seasons because of higher solar radiation (Yoshida and Pardo, 1976; Seshu et al., 1989). The stage of growth is not only important as to the temperature response, but also the CO_2 response of rice. Cock and Yoshida (1973) found that both pre- and postanthesis treatment for 30 days each at 900 ppm of CO_2 increased the yields of grain. Pretreatment increased both grain number and grain weight and posttreatment increased the percentage of filled grains and grain weight.

Rice, being a C_3 plant, is one of the most yield responsive of all plants to elevated levels of atmospheric CO_2. Seldom have the optimal responses in experimental treatments been realized by doubling or tripling the current ambient levels and only during the daylight hours (Baker et al., 1990). The optimal CO_2 concentration for growth and yield lies between 1500 and 2000 ppm (Yoshida, 1976). Furthermore, there are clear sensitivities to both low and high temperature-induced sterility, and varietal differences exist for both temperature sensitivity with respect to both paddy water and air temperatures along with night and day temperatures as to the stage of plant development and time of exposure to elevated levels of atmospheric CO_2.

Recent results from IRRI (1993, 1994) tend to contradict some earlier observations, namely, overall vegetative growth and development were better at 33/25°C, day/night temperature than at higher or lower temperature regimes, and rice benefits from elevated CO_2 are restricted at atmospheric levels up to 500 ppm. Baker and Allen (1993a, b) summarized their detailed studies of temperature sensitivities of rice. Temperatures greater than 35°C for more than 1 h at anthesis induce a high degree of spikelet sterility. Grain yields were highest in the range of 28/21/25°C (daytime/nighttime/paddy water temperatures) at both 330 and 660 ppm of atmospheric CO_2. Rice grain yields declined by about 10% for each 1°C rise in day/night temperatures above 28/21°C.

Farmers of east central China, in the Fujian province with its typical tropical lowlands, subtropical mountains, and hilly areas, have long been familiar with temperature sensitivities of the rice plant. They have wisely laid out models of cropping times, varieties, and growing seasons for rice in what they call climatic layers—the warm, the temperate, and the cool. The warm is suitable for double cropping, mid-maturing, and late-maturing varieties. The temperate climate layer is suitable for single- and double-cropping rice. For single cropping, mid- or late varieties are used. For double cropping, it is mid-maturing cultivars. The cool climate layer is for single cropping and mid- or late-maturing cultivars. Ratooning rice can be planted only in warm and temperate climate layers or zones. Attention should be to avoid the "three coldness" (spring, early summer, and early fall coldness) in arranging double-cropping and ratooning rice. In single cropping of rice, the heading and filling-out periods should be arranged in August and September to have excellent sunlight and temperature (China National Rice Research Institute, 1991).

Rice culture, worldwide, is now limited more by cold than by overly hot temperatures. Shortages of rice in Japan in 1993 were attributed to a stunting of the crop by an unseasonably cold and wet growing season, creating a 1 million ton shortfall. A current primary research effort on the Hokkaido island of Japan and in the Heilongjiang province of China is to develop rice varieties that are less sensitive to cold temperatures and to extend its culture in climates that are currently too cold for successful production. A global warming, independent of the direct beneficial effects of CO_2, might favor world rice production. For 40 centuries, rice farmers in China have endured great variations in growing-season temperatures. These year-to-year variations dwarf any likely change in mean weather variables. Seshu et al. (1989) conclude that variables of a changed climate affecting rice in decreasing order of confidence of visible change would be the atmospheric CO_2 concentrations, air temperatures, solar irradiance, and precipitation. They conclude that in rice breeding, 40 to 50 selection cycles could be completed before atmospheric levels of CO_2 reached 600 ppm. This would provide rice breeders an opportunity to develop varieties with guard cell morphology and temperature tolerances to withstand a CO_2-induced changed environment of temperature and water supplies. Global warming, will, in principle, allow a northward expansion of rice-growing areas, and increases in lengths of growing seasons now constrained by low temperatures. Some rice-based cropping systems would allow for two crops where now only one is possible (Seshu et al. 1989).

Wheat

Wheat is unique as a cereal grain. It is grown on about 250 million hectares, an area larger than any other crop, and is the most widely grown plant in the world today. It contributes more calories and protein to the human diet than any other food. Annual output exceeds 500 million metric tons. As a world trade commodity, it exceeds all other grains combined. Unlike rice, used almost exclusively and directly for human food, wheat is often fed to livestock, especially in Western Europe. Wheat is the most suitable of all grains for bread making, because it contains gluten, an elastic form of protein. When leavened, dough ferments the gluten and traps minute bubbles of CO_2. This causes the dough to rise (Hanson et al., 1982). As standards of living rise, wheat becomes a more important part of the human diet, replacing corn, barley, sorghum, millet, rye and, sometimes, even rice. This is true today in all the major populated countries—China, India, Russia, Indonesia, Japan, and now the Philippines, Europe, and Mexico. For some time, wheat has been the staple food of Russia, Europe, Albania, Bulgaria, Cyprus, Greece, Hungary, Italy, Malta, Portugal, Romania, Spain, Yugoslavia, Algeria, Libya, Iran, Iraq, Israel, Jordan, Lebanon, Saudia Arabia, Yemen, the United States, Canada, France, Turkey, Australia, New Zealand, Chile, Peru, Argentina, Brazil, Egypt and many other countries. *Since 1984, China has become the world's largest wheat producer, now followed by the United States and Russia* (Wittwer et al., 1987).

One might logically project, with the passage of time, that wheat will become increasingly important as water resources for agriculture become increasingly critical. Wheat products include all types of baked goods, pasta, breakfast cereals, alcohol, starch, and straw for animal bedding.

Wheat is a very adaptable species and is grown extensively on every continent. Its climatic adaptiveness is unique among the cereal grains. Of the major food crops, wheat is relatively drought resistant and moderately frost resistant. It can be grown under temperature extremes ranging from –40°C in winter in parts of Canada, Alaska, China, and Russia to +40°C during grain filling in some low latitude regions. It also grows in areas with wide differences in photoperiod and water availability. One of the greatest climatic constraints for wheat is the lowland humid tropics, which comprise 28% of the global land mass. Such areas are adapted for rice culture, but not for wheat. In the lowland tropics, wheat is severely affected by leaf and ear diseases, notably *Helminthosporium sativum*. New sources of heat tolerance and resistances to diseases that frequent the lowland tropics are being sought. With the existing gene pool of bread wheat, sufficient variation in plant development to produce varieties for the lowland tropics has not been found. Major efforts, however, are now in progress in both China and the International Wheat and Maize Development Center in Mexico, D. F. Mexico, to achieve this goal (Saunders, 1991).

Climatic adaptations in wheat relate to two types of wheat. Winter wheat possesses a combination of genes that in temperate climates permits it to be seeded and to germinate in the autumn and to survive winter temperatures as low as –30°C, usually under snow cover, and to later grow, flower, and mature rapidly before the hot, drying summer winds occur. Spring wheat, on the other hand, is a group of wheats that can be sown in any season, if temperature and moisture conditions are suitable.

Battelle (1973), Rosenberg, (1983), Waggoner (1983), and Wittwer (1980) have noted not only that major agricultural crops are found thriving across strong climatic gradients of temperature and moisture, but technology has expanded areas of successful production to make food crops progressively more adapted to extreme climatic conditions. This dynamic nature of adaptation is evidenced by the northward advancement, with no change in southern boundaries, of the production belt for hard red winter wheat in both North America, China, and Russia. Expansion in the United States has been both northward and westward by hundreds of kilometers during the last four decades, as new cultivars and agronomic practices were developed to ensure more stable yields in progressively cooler climates. Thus, the production of winter wheat, with both improved yields and quality, is replacing former areas confined only to spring wheat production (Seshu et al., 1989). Similarly, the development of early maturing cold-resistant hybrid corn varieties has not only dramatically increased yields, but pushed the limits of corn raising in North America 500 miles to the north, during the last 60 years (Battelle, 1973) and furthered extensive plantings in many northern European countries since the early 1960s. Rosenberg (1992a), using climatic analogs, showed that an expansion of the hard red winter wheat zone in the United States (1970–1980) occurred over climatic gradients of temperature and precipitation projected by GCMs, following a doubling of atmospheric CO_2. The yield increase was positive because winter wheat, being more productive, replaced spring wheat.

The wheat plant has an amazing ability to adapt to shortages of water and adverse temperatures. Wheat may be grown at high altitudes in low latitude areas such as in Kenya. However, at low altitudes in the tropics, the growth and yield of wheat are limited, either by long, dry summers, or a high humidity, or by continuously superoptimal temperatures. In hot-dry summers, the life cycle of the crop may be tailored so that grain filling is not prematurely terminated by high temperature and drought (Hoogendoorn, 1985).

Wheat is partially buffered from the effects of drought via several mechanisms (Austin, 1989). In natural environments of wild species, ear emergence and anthesis occur just before temperatures start to increase rapidly and reserves of soil moisture are exhausted. Wheat is also naturally buffered from the effects of late drought by several mechanisms. First, droughts in environments where most wheat is grown become progressively more severe as the season advances. Drought after the formation of tillers is complete will cause those late formed tillers to die and the more severe the drought, the more will die. This will reduce water loss and likely trim supplies to a level likely to be satisfied by what is available. Second, drought advances development (times of ear emergence and anthesis) helping the shoots that survive to escape the worst effects of a late drought. Finally, drought reduces shoot growth and the demand of the vegetative organs for assimilates at a time when photosynthesis rates are near optimal. These assimilates are then mobilized for grain filling.

A detailed simulation study (Rosenberg, 1992a, 1994, Rosenberg and Crosson, 1991) depicting the yield responses of dryland and irrigated corn, wheat, soybeans, and sorghum in Missouri, Iowa, Nebraska, and Kansas (the MINK project), and using a climate analog of the hot-dry climate of 1931 to 1940, with the climate of 1951 to 1986 as the control, found that the yields of wheat were not reduced unlike those of corn, soybeans, and sorghum. If new technologies and the direct effects of atmospheric CO_2 levels of 100 ppm above the current ambient were factored in, yields were significantly increased. Unlike corn, soybeans, and sorghum, the growing season for wheat under the hot-dry climate analog was left unchanged. Under a drier and warmer climate, such as that experienced in the United States during the 1930s, farmers would likely adapt by increasing wheat and sorghum production along the western boundaries of the areas in the United States where corn is now produced (Decker et al., 1985).

The agricultural potential and the possibility of significantly increased yield of harvestable product (harvest index) from future elevated levels of atmospheric CO_2 are of considerable economic and social interest and of unusual potential with wheat. It is the most widely grown and adapted of all food crops. It is a C_3 plant and one of the most responsive, perhaps the most responsive of all cereal grains to high CO_2 levels. This direct response is first reflected in an enhancement of photosynthesis, with resultant significant increases in biomass accumulation, vegetative growth, yield of grains, harvest index, leaf areas, tillering, and with greater root weights and lengths (Cure, 1985; Kimball, 1985; Rose, 1989). Second, it is reflected in the greater water use efficiency of wheat at elevated levels of atmospheric CO_2. Of all the major food crops observed, the yield increases from high CO_2 for drought stressed wheat plants was the greatest (Gifford, 1977, 1979a, b; Kimball, 1985; Sionit et al., 1980, 1981). Penning de Vries et al. (1989) simulated the comparative yields of wheat and rice in current and future climates when ambient CO_2 doubles. This was done for four geographic regions characterized by widely different climatic regimes—a temperate (Netherlands), Mediterranean (Israel), semiarid (Hydrabad, India), and subhumid tropical (The Philippines). Taken together, 25 to 50% increases of average potential yields of wheat and rice were projected for the tropics under future climates. For cooler climates the effects were smaller. A major component of the yield increases under the future high CO_2 climates was the much more efficient use of water resulting from the higher CO_2 concentrations. The effects on yields of both crops would be smaller, with a severe constraint in nutrient availability. Much of the positive response of wheat is the resistance to drought imposed by higher levels of atmospheric CO_2. This is particularly significant in that a major part of the world's wheat is grown in marginal climatic areas (United States, Canada, Russia, India, Pakistan, China, Australia, Turkey) where water is the limiting factor for increases in yield and crop productivity.

Gifford (1989) makes a strong point, with wheat in mind, that the global increases of atmospheric CO_2 represent an improving component of the fitness of the earth's environment for food production. Because yield increase percentages in response to high CO_2 are larger for drought and salt stressed plants than for nonstressed plants, some marginal cropping sites may show less year-to-year variation. This would be an improvement in the stability of food production as well as the magnitude of output, not only for Australia, but for Russia, the United States, northern Europe, India, Canada, Turkey, and Argentina—all major wheat producers. In the Netherlands, for example, with spring wheat, combinations of a temperature rise and higher CO_2 resulted in large increases in dry years, and small increases when water was not limiting growth (Nonhebel, 1993).

For China, the number one rice and wheat producer of the world, Jin et al. (1992) report that the moisture condition is the dominant environmental factor affecting wheat production. When both direct and indirect effects of increased CO_2 on wheat yield were considered, the overall output of winter wheat in China, which represents 85% of the total wheat produced (Wittwer et al., 1987), would increase by nearly 16%, most of which would come from the North China Plains (Jin et al., 1992). From the climate change scenario, derived from a modified GISS model and a projected CO_2-induced warming, the accumulated temperatures during the various growing seasons would obviously increase. There would, accordingly, be profound effects on cropping indices and systems. They could change from a single crop per year at present to three crops in 2 years in the northern subregion, and from two crops per year to dryland triple-maturity in the North China Plains, and the growth pattern of "wheat plus rice" in those areas where water is available. In the middle and lower Yangxi River valley, "wheat plus double rice" and triple rice would likely be the major growth patterns. In the Sichuan province, double or triple rice could be grown in the basins. The conclusions of Wang et al. (1991a) support those of Jin et al. (1992), which are that the impact of higher levels of atmospheric CO_2 will be more significant for wheat than for rice, and that agricultural production in northern China will become more favorable than in the south, with little change in the southern agricultural regions.

Overall, studies of elevated atmospheric levels of CO_2 on wheat in the United States have shown increases in seed yields ranging from 15 to 50%. These increases were in both seed size (weight) and number (Fischer and Aguilar, 1976; Krenzer and Moss, 1975) and for both Mexican dwarf spring wheat and hard red spring wheat. Winter wheat was grown in open-top chambers in the field during winter-time in Delaware (Havelka et al., 1984) over a 2-year period (1979–1980). The plants were subject to CO_2 atmosphere levels of 340 and 1200 ppm during daylight hours at different stages of development. Highly significant yield increases (17%) were achieved when applied from jointing to anthesis and from jointing to physiological maturity. When the extra CO_2 was applied only during seed growth (from anthesis to physiological maturity), the increases in seed growth were minimal. There was no change in harvest index from the CO_2 treatment.

Maize (Corn)

Maize, or corn, is grown in more diverse areas of the earth than any other crop (Abelson, 1989b). Climatically, it is successfully produced in the lowland humid, and from the intermediate hill country to the highlands of the tropics. It flourishes throughout the temperate zone and is found as extensive plantings in the far northeastern parts of China in the Heilongjiang province and at high elevations of the south in the Yunnan province, and in France, with extensive use of plastic soil mulches to hasten its growth. It is found as far north as Alaska in North America, and Scandinavia in Europe. Within a generation, after it had been discovered in the new world in 1492, it had spread throughout Europe and, within two generations, it was being grown around the world in every region suitable for its cultivation. There was no crop equal to its climatic and cultural adaptability (Mangelsdorf, 1986). Maize or corn can be grown under a broader set of environmental conditions than wheat, and it is particularly adaptive to the tropics where wheat is not (Abelson, 1982). In the developing world, it is a major staple food, where half the crop is consumed directly by humans. Acceptability of the high lysine component, which greatly increases the biological value of the protein, is now a reality. In Brazil, high yielding corn can be grown in the toxic aluminum soils of the vast Cerrado plateau, of 200 million ha, and on similar soils in many countries of Africa and the islands of Indonesia. It has become a major crop in Thailand, Europe, Argentina, South Africa, Brazil, India, Indonesia, the Ukraine, and what was Yugoslavia. It is the most important staple food for people in Costa Rica, El Salvador, Guatemala, Haiti, Honduras, Mexico, Nicaragua, Paraguay, Venezuela, Benin, Kenya, Malawi, Somalia, South Africa, Tanzania, Zambia, and Zimbabwe. Early hybrid varieties and improved culture have moved it 500 miles further north in the United States during the past 50 years (Battelle, 1973), and it has become a major crop in Western Europe since the early 1960s. *The United States is by far the largest producer, with half the world's output. Few places are as well-suited for corn as the U.S. agricultural belt, where yields are often three times greater than in the developing world. China is second in world production. Of the major food crops of the world, it is among the highest in photosynthetic efficiency, withstands the hottest of climatic conditions, requires less water than rice, and will produce more total digestible nutrients per unit land area for people and livestock than any other crop.* As a species, corn shows extremely wide adaptation. It is grown as a commercial grain crop from about 55°N to 40°S, and from sea level to 4000 m altitude (Fisher and Palmer, 1983). Genetic and cultural improvements for greater climatic adaptability continue, with two international agriculture research centers (the International Wheat and Maize Development Center in Mexico, D. F. Mexico, and the International Institute for Tropical Agriculture in Nigeria, Ibadan) devoted to its research. It is believed by some that, eventually, production of corn will outstrip wheat production worldwide (Abelson, 1982).

Corn is a major C_4 food crop of both the agriculturally developed and less developed world. Its photosynthetic metabolism differs from that of the two other major cereals, wheat and rice. Associated with the C_4 photosynthesis is an extremely low rate of photorespiration, distinctive leaf anatomy, a low CO_2 compensation point, absence of photosynthetic light saturation up to full sunlight, and better adaptations to high-temperature and high-insolation conditions. Although there is great plasticity within the species, the range of adaptation for individual cultivars is very limited. As with rice and wheat, the potential effects of both climate change and CO_2 enrichment on plant growth and productivity have been studied. Thus far, however, much greater emphasis has been placed on the effects of climate change than on the direct effects of CO_2 enrichment. More climatic data are available from real world experience than for any other crop, while little or no data exist outside of growth chambers or greenhouse experiments on the crop's response to an elevated atmospheric level of CO_2. With the warmings and accompanying dryings predicted by some climatic scenarios, Blasing and Solomon (1982) predict that the western portion of the United States corn belt will be unable to sustain production except for irrigation. Similarly, for every 1°C increase in regional temperature, a northeastern movement of 175 km for the eastern part of the United States corn belt has been predicted by Newman (1982). Decker et al. (1985) also predicted that if summers were to become drier and hotter, maize production in the western United States corn belt would be reduced in yield so much that it would be replaced with sorghum and millet. Corn is unusually sensitive to drought at silking and during early grain development. This was evident in yield depressions wrought following the drought in the United States in 1983 and again in 1988.

With corn being a C_4 plant, and unlike most C_3 plants, the greater effect of elevated levels of atmospheric CO_2 is on water use efficiency, rather than increases in net photosynthesis. Water use efficiency is the amount of carbon or dry matter accumulated per unit of water taken up or lost by the plant. The change in water use efficiency comes from changes in both photosynthetic and transpiration rates. For corn, the predominant effect is from transpiration; for C_3 plants, such as rice, wheat, cotton, and soybeans, the primary effect is from photosynthesis (Acock, 1990). Corn, with its C_4 pathway for photosynthesis, traps ambient CO_2 quite efficiently (Rose, 1989). This has led Acock (1990) to suggest that corn may eventually lose its photosynthetic advantage over rice and wheat and other C_3 plants as the atmospheric CO_2 levels continue to rise, and we may see less corn produced.

Despite the global significance of corn as a world food crop and to American agriculture, little field work has been done on its yield and dry matter response to elevated levels of atmospheric CO_2, and the results are not consistent. Surano and Shinn (1984) found no significant increases in photosynthesis or on phenology, but did report a significant increase in water use efficiency for field corn grown under two water regimes, the greater being with water-stressed plants, but a more rapid emergence of leaves was noted by King and Greer (1986). Moreover, Carlson and Bazzaz (1980) reported a 26% increase in the total plant weight at 600 ppm of CO_2, and Rogers et al. (1983a, b) observed a 60% rise in dry matter over ambient CO_2 concentration. Others found no significant differences (Carlson and Bazzaz, 1980; King and Greer, 1986). Acock and Allen (1985), in a review of a report by Rogers et al. (1983a, b), emphasized that elevated levels of atmospheric CO_2 significantly increased the partitioning of dry matter to roots of 11-week-old corn and soybean plants.

Other Cereal Grains

Other cereal grains as world food resources include grain sorghum, pearl millet, barley, oats, rye, and Ethiopian teff or "t'eff."

Millet and sorghum are the staple foods of 12 countries in an area extending from West to East Africa and on into the Arabian Peninsula. With a combined population of about 250 million these countries include the Cameroon, Chad, Mali, Mauritania, Niger, Nigeria, Senegal, the Sudan, Togo, Uganda, Yemen, and Burkina Faso. Noting the significant contributions of these cereals to the human diet, and that both sorghum and millet are C_4 plants and some of the most resistant of all major food crops to hot-dry climates, it is indeed surprising that so little is known concerning either the potential effects of climate change or CO_2 enrichment on yields and productivity. Sorghum and millet are the most resistant and tolerant of all cereal grains to high temperatures and deficiencies of precipitation, with the possible exception of teff, a cereal grass (MacKenzie, 1985). Millet can be grown in warm, moist climates as well as in rather cool, dry areas. The many varieties are very adaptable to unfavorable growing conditions, such as infertile soil, intense heat, and scanty rainfall. Millet can tolerate drought because of strong roots that penetrate deep into the soil. Sorghum needs a climate similar to maize or corn. Optimal temperatures for photosynthesis are close to 40°C, but optimal growth temperatures are generally 30 to 35°C. The temperature adaptation range is immense with some cultivars doing well at near 40°C and some at less than 25°C (Eastin, 1980). But millet can grow in a wider range of climatic conditions than corn, needs less water, and is relatively drought resistant. Reference is made that with the projected limitations of water for agricultural purposes, irrespective of climate change, sorghum and millet could replace corn and wheat and even rice in many agricultural areas. Of the 7% of the world's arable area devoted to the production of sorghum and millet, 70 million hectares are concentrated in the less developed economies of Asia and sub-Saharan Africa, and production is climate dependent (Rao et al., 1989). Research is currently directed toward cultivars of sorghum and millet that are even more resistant to drought, and with greater yield stability, and adapted to the dryland agriculture in the semiarid tropics of Asia and Africa (Rao et al., 1989). Moreover, sorghum and millet are the most important cereals for resource-poor farmers, especially in West Africa. They are the only staple food crops that can withstand the ravages of temporal and spacial variations in rainfall, and rainfall shortages, that frequently occur (Sivakumar, 1989). Teff (*Eragrostis abyssinica),* a cereal grass, thrives in the semiarid highlands of central Africa, and is specifically grown as a staple food crop for making a flat bread called "injera" in Ethiopia. Teff needs less rain than corn, millet, or sorghum, and, in Ethiopia, can be grown in higher and colder regions than either. It needs less rain than high altitude and latitude crops of wheat and barley. It can better withstand the occasional floods that occur and roots can respire with minimal oxygen.

Sorghum and millet are especially important in China, with production stretching from 32 to 50°N latitude and from 108 to 130°E longitude. Together, they occupy about 5% of the total arable land of that country. These two crops are highly resistant to the climatic hazards of drought and floods. Millet is a main food in the northern part of China. It will tolerate shortages of water better than any other major food crop. As two of the most ancient of Chinese crops, millet and sorghum function as food security when other crops (wheat, corn, soybean, and rice) in China fail.

Oats, barley, and *rye,* by contrast, are adapted to the temperate zones of the northern and southern hemispheres and the highlands in the tropics. They are especially abundant and productive in northern Europe.

Barley deserves special note. Among all the cereal grains, it has the shortest vegetative period, is the earliest in maturity, and most resistant to hostile environments. Canada and the United States with their 3.9 and 3.0 million ha and annual outputs of about 11 and 9 million metric tons, respectively, are the world's leading producers. Some varieties are highly tolerant to low temperatures and salty soils. It may be planted in the fall and harvested the following spring, or unlike any other cereal grain, may be planted in one autumn and harvested the following autumn, as in Tibet. Naked seeded barley, grown at elevations up to 4200 m, is the main food crop in the Xizang Plateau of Tibet, with extensive plantings in Inner Mongolia (Wittwer et al., 1987). Barley is successfully grown in all northern Scandinavian countries, the British Isles, and as far north as Fairbanks, Alaska.

Oats are grown extensively in such diverse places as the United States, Russia, Canada, Sweden, China, Poland, France, Germany, Finland, and Norway. Rye is found principally in Russia, Poland, and Germany.

Experiments exclusively on the effects of elevated levels of atmospheric CO_2 on the productivity of grain sorghum, pearl millet, oats, rye, and barley have not been done (Acock and Allen, 1985; Cure, 1985; Palutikof et al., 1984). With grain sorghum and pearl millet as C_4 crops, a pattern of response comparable to that of corn could be expected. For barley, oats, and rye, some similarity to the response of wheat would likely follow. This has been verified by Krupa and Kickert (1993) for oats and barley, and Pettersson et al. (1993) for barley. The positive responses from elevated atmospheric CO_2 were increased tillering, leading to more ears per plant, increased harvest index, and mean kernel weight. Substantial increases in root dry weights of sorghum have been reported when grown in an elevated level of atmospheric CO_2. Plants had higher root numbers and dry weights at all soil profile depths (Chaudhuri et al., 1986).

Tuber and Root Crops

Potato

The fourth most important world food crop is the potato. It is a product of South America and had its origin in the highlands of Peru and Chile (McNeill, 1991). It was introduced to Spain and England in the 16th century, and from there spread to other European countries. Gradually, the potato became one of the main food crops of the world. During the 19th century, one nation—Ireland—became so dependent on it as a staple food that it was called the Irish potato. When late blight (*Phytophthora infestans)* hit in 1846, one of the world's greatest famines resulted. More than 1.5 million Irish died and another 1.25 million quarter migrated. The Irish potato famine exemplified the interaction of climate (wet-cool seasons and torrential rains) and pests in devastating a crop and a people (National Research Council, 1972; National Academy of Sciences, 1976).

Of the approximately 4 billion tons of food produced on earth each year, the potato, now produced on 50 million ha of cropland worldwide, accounts for over 300 million tons, ranking just behind wheat, corn, and rice, and comparable in nutrient value, providing an excellent source of calories and protein, plus vitamin C not found in cereal grains. While the cultivated potato is grown all over the world, it is found most prominently in western, northern, and eastern Europe, particularly in Russia, Poland, and Germany, where it is a leading food crop. There are over 9 million ha in China, with nearly an equal area in India. An international agricultural research center, devoted to genetic and cultural improvements, is located near Lima, Peru. The potato is climatically adapted and grown successfully at some locations or seasons of the year in most all countries in Asia, Africa, Latin America, North America, Europe, and every state in the United States, including Alaska and all of the provinces of Canada. The highest average yields are obtained in countries with a moderate climate where there is a day length of 13 to 17 hours during the growing season, and an average temperature of around 15 to 18°C, and where rainfall or irrigation provides an ample water supply.

The potato may be found growing successfully as a food crop in every province, autonomous region, and municipality of China, extending from the most northerly to those in the far south. It is grown as a winter crop in India's Punjab, in rotation with rice, wheat, and corn, in the rice paddies of the highlands of Sri Lanka, alternating with rice, the Nile valley of Egypt, and on higher elevations of Kenya and Ethiopia. It is adapted to mature within a growing season of less than 90 days and is one of the most important food crops for the state of Alaska. The potato, in summary, is adapted for successful production in such a great number of climatic regions of temperature, sunlight, and photoperiod and alternate growing sites, it is difficult to visualize that a global warming, even as projected by some of the most extremes of scenarios, would have a major negative world impact.

Again, CO_2 enrichment for potatoes is fragmentary, but what has been reported is conclusive. The potato as a C_3 plant, is very responsive to elevated levels of atmospheric CO_2. The early field work of Chapman and Loomis (1953) and Chapman et al. (1954) showed that a doubling of the CO_2 content of the air doubled photosynthesis, and that photosynthesis was increased fourfold by increasing the CO_2 content to 1500 ppm. Moreover, it appears that in the potato, the products of increased photosynthesis are used almost exclusively by the developing tubers, which apparently serve as a sink (Arteca et al., 1979; Collins, 1976; Wheeler et al., 1991). When CO_2 enrichment of the root zone occurred, there was a very dramatic (severalfold) increase in tuber growth. The effect was not so much on tuber number but on tuber size and weight. It is truly unfortunate, with the potato so widely grown worldwide as a major food staple for both developed and less developed countries, that so little research has been done on its response to continuously elevated levels of atmospheric CO_2, and, more particularly, with the effects of the climatic variables of temperature, sunlight, soil nutrients, and water factored in.

Sweet Potato

Root crops, of all the major food crops, are by far the best adapted to the lowland humid tropics. High rainfall and humidity often adversely affect the reproduction, ripening, drying, and storage of cereals and grain legumes and increase the severity of pest and disease problems, but not for sweet potatoes, cassava, yams, taro, and cocoyams (Pendleton and Lawson, 1989). There are no places on earth that are too hot or humid to grow these crops; and some, such as the cassava, are extremely resistant to drought and soils of low fertility, low pH, or high levels of aluminum that would be toxic to other crops.

Sweet potatoes are grown over a wide range of environmental conditions between 40°N and 40°S latitude, and from sea level to 2300 m altitude. It is a perennial plant but is treated as an annual with a normal growing season of 3 to 7 months. Although it is grown in relatively high rainfall areas, it has good drought tolerance. Two thirds of the world production is in Asia. Low night temperature (20°C) and long days favor tuberous root initiation and development. Tuberous root development was more rapid at night temperatures of 25 than at 30°C, but no tuberous roots formed at 10°C. When sweet potato plants in the field were exposed to drought, tuberous roots developed slowly but resumed growth when stress was removed (Hahn and Hozyo, 1983).

The sweet potato, like the Irish potato, originated in South America. It is a primary, as well as a secondary source of food for hundreds of millions of people in many countries that are a part of Africa; South, Central, and North America, and Asia. Extensive plantings are found in Kenya, Uganda, and Nigeria, which have more reliable food supplies than Somalia, Ethiopia, and the Sudan. It does best in rain-fed areas. It is an ideal crop for low-income farmers and has been promoted for such, even in the United States, because it grows well and is productive in low nitrogen soils, tolerates drought, crowds out weeds, and suffers from few pests. As a rich source of carbohydrates, it also has a high carotene content, with genetic selections that are increasingly higher in nutritive values. It can be harvested for food at various stages of maturity, where grain crops fail. The sweet potato can also be stored in the ground for several months and harvested as needed for food. China is the world's largest producer of sweet potatoes (Wittwer et al., 1987), where both the roots and leaves are used to feed livestock, and as a food for people. This staple food crop has saved the Chinese on many occasions from the ravages of famine when grain crops were in short supply. It is grown both as an intercrop with corn to prevent erosion and improve soil stabilization on the many steep hillsides of Sichuan and alongside rice in the flat plain areas of Anhui provinces in China. Freshly baked sweet potatoes are relished particularly in China, where they are sold as a fast food on the streets of the major cities and towns. They require much less water than rice. Two international agricultural research centers—the Asian Vegetable Research Development Center (AVRDC) in Taiwan and the International Institute for Tropical Research (IITA) in Nigeria—direct resources for research on the sweet potato. Varieties have been developed that mature in 3.5 months and will yield up to 15 tons/ha without added fertilizer, and around 30 tons/ha with fertilizer.

Cassava

Cassava is the common name for *Manihot esculenta.* Manioc is the name for it in Latin America. As a granulated product, it is known as tapioca in the United States (Kahn, 1985; Wigg, 1994). Cassava is grown throughout the lowland tropics on sites not suitable for rice, in the Americas, the Caribbean, Africa, and Asia, including southern China and India. It is the fourth most important food staple within the tropics as a source of calories for people, after rice, maize, and sugarcane. It is the main sustenance today for some 500 million inhabitants of many countries in Africa, Asia, and South and Central America (Cock, 1982). It is widely grown in Thailand, Nigeria, Zaire, Angola, Burundi, Rwanda, the Congo, Cabon, Ghana, Mozambique, Tanzania, Colombia, Brazil, Paraguay, Taiwan, and the southern provinces of China, including Tainan Island. In these countries cassava is a basic energy food and animal feed, has many industrial uses, with the potential for more, and is produced on marginal agricultural lands of low fertility and low moisture, and often with no fertilizer added. It is extremely tolerant of soil acidity and associated high aluminum levels. Excellent crops can be grown on such soils in the tropics. Soybeans grown in the same soil would fail.

Cassava is very resistant to locust attacks and drought. It can easily be grown on almost any kind of soil in warm climates, and in regions having a long dry season and where irrigation is not feasible. Once it has achieved its full subterranean spread, a single plant will produce 50 lb of food, or 25 tons/ha as a worldwide average. Under good conditions, it will produce five times that amount. Harvest can be initiated within 3 to 6 months of planting, and can be continued for up to 18 to 20 months after it matures. It can be left in the ground for up to 2 years, providing a local food reserve. Cassava, similar to the sweet potato, can be grown as a famine reserve crop. It has a remarkably high harvest index, of up to 81%.

Cassava is grown between latitudes 30°N and 30°S in areas where the annual rainfall is greater than 750 ml and the annual mean temperature is over 18°C. Small plantings may be grown near the equator in South America and Africa, at altitudes of up to 2000 m with annual mean temperature as low as 17°C if there are minimal seasonal fluctuations (Cock, 1982). No place on earth is too hot for the growth of cassava if there is water. Its chief climatic constraint are temperatures that are high enough for its growth. It is an ideal crop for the tropics, with cheap calories from the tubers, good quality protein from the leaves, and large quantities of nectar from the flowers (Hahn, 1993).

In summary, significant cassava production is found in six important subclimate areas: the lowland humid subtropics, the low subhumid tropics, the lowland and highland semi-arid tropics, the lowland hot savannas, and the highland humid tropics (El-Sharkaway, 1993).

Drought is one of the major agricultural disruptions prominently portrayed to occur with global warming. Cassava is uniquely drought resistant and tolerant of prolonged drought (Cock, 1982; El-Sharkaway, 1993). During drought stress cassava follows a conservative pattern of water use by closing its stomata and reducing the formation of new leaves. The leaves that remain on the plant have a remarkable ability to actively photosynthesize when moisture again becomes available. Thus, the plant slows its growth during drought periods, but rapidly recovers when the rains come. Unlike many other crops, cassava, once established, has no critical stage or phenological period when drought will cause a disastrous decrease in yield, which is true of all cereal grains, legumes, pulses, and most fruits and vegetables. Hence, cassava is well adapted to areas that experience a long, dry season or uncertain rainfall typical of the semiarid tropics (Cock, 1982; El-Sharkaway, 1993).

Total biological yield ultimately will depend on the efficiency with which a crop can convert a finite water supply to dry matter. To this end, the cassava is the most meritorious. It is highly productive under favorable environments where no major production constraint, particularly water, prevails. Unlike other staple food crops, such as cereals and most legumes, cassava can produce reasonably well with virtually no purchased inputs. The slow, early establishment of cassava makes it possible to intercrop it with crops having a short growth cycle, such as beans or cowpeas, with minimal competition and yield loss. It is more efficient to grow cassava intercropped with such legumes than to grow the root crop and legumes separately.

Cassava is a major world food crop, with an annual global production of approximately 160 million tons (1991), and is used in various ways. In Paraguay, the per capita consumption is nearly a pound per day, providing more than 11% of the national caloric intake. In Brazil, which produces about 19% of the world's crop, most people use it in a roasted-flour form complimented with rice and beans. Nineteen million tropical acres are devoted to its production in the countries of Africa, primarily Nigeria, Zaire, and Tanzania (El-Sharkaway, 1993; Kahn, 1985). Nigeria has recently become the largest cassava producer in the world (Hahn, 1993). Among the latest developments at IITA is a tetrapoloid "supercassava" that has yielded up to 70 tons/ha (Wigg, 1994).

Yams and Cocoyams

Yams and cocoyams (taro) represent two other important tropical root crops. Yams are adapted to areas of fairly high rainfall and with a distinct dry season of not more than 5 months. Because their growing period exceeds 5 months, yams are not grown in areas with rainy periods of less than 5 months. They have an annual growth cycle of 5 to 10 months and dormancy for 2 to 4 months. Nigeria is the world's leading producer of yams, where its contribution to the human diet is comparable to that of cassava (Hahn and Hozyo, 1983). Cocoyams (taro) are a less important root crop grown primarily in the tropics of Oceana and Africa. In Nigeria, Africa's chief producer, the quantity is less than one eighth that of yams and about one fifth that of cassava (Nweke, 1987).

The effects of an enriched level of atmospheric CO_2 on cassava and sweet potato are remarkable. Perhaps the most pronounced and universal result is the enhancement of root growth. This is good because the root is precisely that part of the plant used for human food. The effects are manifest in the root system architecture, micromorphology and physiology (Rogers et al., 1986, 1992a, b), and root/top ratios. Root lengths and dry weights of roots are dramatically increased. The immediate and long-term implications with respect to all food crops under drought—their survival, utilization of mineral nutrient and water resources, and especially the magnitude of food production with particular reference to root crops, where the harvest index is the root, are enormous. As might be expected both cassava and sweet potato, like the Irish potato, show a remarkable response in yield enhancement of roots to elevated levels of CO_2, especially if accompanied by increasingly higher temperatures. The yield responses exceed the averages for cereals and other food crops—possibly because of their relatively simple storage forms. As has already been indicated, elevated levels of atmospheric CO_2 increase the temperature optima for photosynthesis (Mooney et al., 1991). Idso (1989a) demonstrated that with carrot and radish, two cool season root crops, the effects of added CO_2 on root growth may be minimal at relatively cool temperatures, but become greater and greater as the temperature rises. The response of carrot greatly exceeded that of radish. With the radish, however, Wong (1993) found that using variables of atmospheric CO_2 at 350 and 700 ppm and relative humidities of 35 and 90% at 32°C during the day, storage root mass was doubled at 700 ppm CO_2 at high humidity and showed a 3.5-fold increase at low humidity.

The detailed studies of Bhattacharya et al. (1985a, b, 1990) on the effects of elevated atmospheric levels of CO_2 on the growth and yield of the sweet potato are revealing. Sweet potatoes, grown from cuttings, maintained in growth chambers, and subjected to continuous atmospheric levels of 350, 675, and 1000 ppm of CO_2 resulted in a significant partitioning of biomass into the roots. Both the number and the size of roots increased with a marked increase in root/shoot ratios. Tuber growth at 675 ppm was 1.8 times that of 350 ppm, and 2.6 times at the 1000 ppm CO_2 level. Root growth in stem cuttings exposed to similar levels of elevated CO_2 was also significantly enhanced (Bhattacharya et al., 1985b). Later studies by Bhattacharya et al. (1990) with sweet potatoes grown in open-top chambers in the field exposed to varying levels of CO_2 (364, 438, and 666 ppm) resulted in increases of root/shoot ratios, yields of fresh storage roots, and storage root starch even under water stress conditions. They concluded that CO_2 enrichment not only alleviated the detrimental effects of water stress, but also allowed the stressed plants to maintain essential metabolic processes that are otherwise disrupted in water-stressed plants.

Equally significant, and perhaps even more striking, have been the studies of Imai et al. (1984) on cassava (*Manihot esculenta* Crantz) and Imai and Coleman (1983) on Konjak (*Amorphophallus konjac*, K. Koch). Konjak is a common root food crop grown in Japan, which produces large tuberous corms used for making flour.

Cassava plants were grown continuously for 2 to 3 months at 350 and 700 ppm of atmospheric CO_2 at temperatures of 35°/26°C day/night. There was a 150% increase in total dry matter with a doubling of CO_2. The most vigorous growth occurred under high temperatures and with high CO_2. The lack of further information currently available as to the response, not only of the potato, sweet potato, and cassava, but other major tuber and root staple crops (yam, cocoyam, taro, sugar beet) is appalling. No studies have been reported on the response of either yams or cocoyams to elevated levels of atmospheric CO_2. This is when one considers their importance as some of the major food staples of the developing world, and especially their promise for low input production on marginal soils with minimal water resources prevalent in so many agricultural developing countries in the tropics and subtropical region.

The Legumes

The legumes—peas, beans, pulses, peanuts (groundnuts), lentils, and lupines—hold a key position in most food-producing systems, for all nations and on all continents. They are a direct food resource providing 20% of the world's protein for human consumption, and one fourth of the world's fats and oils. The soybean alone provides about two thirds of the world's protein concentrate for livestock feeding, and is a valuable ingredient in formulated feeds for poultry and fish. The soybean also provides about three fourths of the world's trade in high protein meals. Legumes are also valued as soil-enriching sources of green manures, and in crop rotations and as cover crops. They are an excellent source of protein in human nutrition and a supplement for the cereal grains. Additionally, legumes have the unique capacity of participating in symbiotic relationships with nitrogen-fixing bacteria, having the ability to capture atmospheric nitrogen to make it available for crop production. Moreover, legumes are climatically adapted to a great variety of temperatures, daylengths, and moisture regimes.

Soybean

Of all the legumes, the soybean is best known for its climatic adaptations and limitations. It has probably received far more attention than all other grain legumes combined. The environmental adaptation of the soybean is similar to that of corn. It is grown from latitudes of 0 to 55°N. Major production is between 25 and 45° N and S latitudes at altitudes of less than 1000 m. The soybean is temperature sensitive and is usually grown in environments with temperatures between 10 and 40°C during the growing season (Wigham, 1980). Successful production may be observed from the most southerly to the most northern provinces in China, where it had its origin. In China, soybeans have been classified into 13 maturity groups, correlated with both latitude and elevation. In the United States, which produces over half the world's supply, they are produced from Minnesota in the north to Mississippi, Alabama, Louisiana, and Texas in the south. Soybeans are grown successfully throughout India, Indonesia, Paraguay, and in many parts of Brazil and Argentina, and many equatorial countries in Central and South America and Africa. In the tropics of Indonesia and Nigeria, maize and soybeans go together. Where one grows, so does the other. This is also typical of the U.S. corn belt, and many of the central provinces in China. Their culture is extensive in the state of São Paulo and elsewhere in Brazil, which is second only to the United States in world production. It is difficult to believe that a degree or two of global temperature increase or decrease would significantly alter world production of this significant crop.

Other Seed Legumes

Concerning the response of legumes to elevated levels of atmospheric CO_2, the soybean has been the prototype for favorable results. Atmospheres enriched with CO_2 promote greater root growth as expressed by root/shoot ratios and remarkable increases in the weights of roots, root lengths, root, stele diameters and cortex widths, leaf and stem weights, and total plant dry weights (Rogers et al., 1992a). Root growth is enhanced much more than that of shoots. Surprisingly, however, the soybean appears to be the only crop species for which additional atmospheric CO_2 usually decreases the harvest index (Cure, 1985; Rogers et al., 1986). The response to CO_2 increases with each increment of CO_2, and, unlike cereals and grasses, the nitrogen content of various plant parts does vary significantly with CO_2 enrichment (Reddy et al., 1989).

In addition to positive effects on plant growth, there is a striking impact on biological nitrogen fixation. The amounts of nitrogen fixed by leguminous plants subjected to CO_2 enrichment are truly dramatic. Severalfold increments in biological nitrogen fixation have been observed from increases in atmospheric concentrations of CO_2. A doubling of the atmospheric CO_2 has doubled the amount of nitrogen fixed (Hardy and Havelka, 1977). Other studies have shown substantial, but more modest increases (Cooper and Brun, 1967; Hardman and Brun, 1971; Rogers et al., 1981; Williams et al., 1981). Thus, Kimball (1985) has surmised that leguminous crops will probably not be limited by a lack of nitrogen in the future high CO_2 world any more than they are now.

This could be extended to plants with nitrogen-fixing symbiotic relationships beyond the seed legumes. These include the nitrogen-fixing nodulated trees such as the legume *Robina* (locust) species and the desert mesquite, the actinorrhizal species of *Alnus glutinosa* (red alder), and *Eleagnus augustifolia.* With these trees, substantial increases in plant growth occurred with CO_2 enrichment, despite likely soil nitrogen and phosphorus deficiencies. Plants were both larger and had more nodule mass than the plants in ambient CO_2 (Felker and Bandurski, 1979; Norby, 1987). Still other symbiotic relationships exist with *Anabaena-Azolla* combinations that enrich rice paddies, and the *Azotobacter* and *Spirillum* rhizosphere association in grasses, cereal grains, and other nonlegumes. The best plants are those that excrete carbon compounds to supply energy for nitrogen-fixing bacteria associated with plant roots. A likely improved source comes from CO_2 enrichment, since there is an increase in energy from greater photosynthetic activity (Egli et al., 1970; Stulen and den Hertog, 1993). Poorter (1993) concludes that for nitrogen-fixing plants, and the majority of experiments with CO_2 enrichment, the increase in nitrogen fixation has resulted from an increase in nodule mass, either by an increase in the number of nodules, facilitated by an increase in root size and number of sites for nodule establishment, or by a change in mass per nodule. Poorter (1993) further concludes that CO_2 enrichment is probably favorable for plant–mycorrhizal associations in that more carbon is allocated toward the symbiont, but further experiments are required to establish the extent of stimulation by CO_2 enrichment.

Virtually all investigators dealing with the direct effects of high levels of atmospheric CO_2 have concluded that the impact on food crops will be positive. There are those, however, that emphasize the few exceptions. Even the most vocal critics (Bazzaz and Fajer, 1992), however, admit with legumes, particularly the soybean, the crop will be positively impacted by elevated levels of atmospheric CO_2 and this will include the impact on ecosystem-wide changes, as well (Stulen and den Hertog, 1993).

Surprisingly, little is known concerning either climatic adaptations or responses to elevated levels of atmospheric CO_2 for legumes, other than the soybean. Chickpeas, pigeonpeas, cowpeas, dry field beans or Chile beans (*Phaseolus vulgaris),* mung beans, urd beans, broad ("horse") beans, peanuts, and other leguminous crops are also major vegetable protein food crops in temperate, subtemperate, and tropical regions. This is particularly true for millions of people and livestock in many developing countries in Asia, Africa, and Central and South America. Climatic limitations and a lack of technological inputs have resulted in relatively low productivity and have led to sharp declines in legume-to-cereal production ratios, causing an almost 50% drop in per capita availability during the last quarter of a century. Legumes (pulse crops) in subtropical, semiarid, and tropical regions (with the possible exception of the soybean and some peas) have had a relatively short history of genetic improvement, compared with wheat, rice, and corn. Most are scarcely above the primitive lines. The various varieties (cultivars) within the same crop, as with the soybean, show marked differences in responses to photoperiod, temperatures, and water availability or soil moisture. Many varieties are location specific. Crop stands, disease and insect infestations, and pod set are affected by water availability, humidity, and temperature. Yields of these legumes do not respond favorably to either low or high levels of available water, and generally not to irrigation. Most legumes are rain fed (Khanna-Chopra and Sinha, 1989). Irrigation usually results in more dry matter production rather than increased yields of seed legumes.

Field Bean

The common field bean (*Phaseolus vulgaris* L.) is an important world food crop and is the most widely grown of the four cultivated species of *Phaseolus* from the American tropics. It is the world's most important grain legume for human consumption. Of the mean annual world production (1976–1978) of 8.3 million metric tons, 47% was produced in Latin America and the Caribbean, 16% in Africa, 15% in China, 11% in North America, and 8% in Europe (Laing et al., 1983). Field bean production statistics show a remarkable concentration of production with 76% in microregions in Latin America, with mean temperatures during flowering of 19 to 23°C; the minimum, optimum, and maximum mean temperatures suggested for five locations with 250 genotypes for yields were 12.2, 20.6, and 29.1°C, respectively. Water stress is also critical. The field bean, having a very short growth cycle, is particularly susceptible to short stress periods during the flowering and early pod determination phases. There are also temperature and photoperiod interactions, particularly night temperature, which modify flowering responses in individual genotypes. The extreme sensitivity of field beans to both water stress and temperatures, relating to flowering and seed development, should result in them being among the more vulnerable of food crops to any climatic change. Varieties, however, differing greatly in their climatic responses and adaptations would probably be feasible.

Other than the soybean and field beans, chickpeas and pigeonpeas predominate in world pulse production. Their primary production sites, along with mung and urd beans, are the Far East and Southeast Asia, where chickpeas constitute 38%, and pigeonpea 20%, of the world's total. Pigeonpea is a crop grown primarily in India, which accounts for 77% of world production. In India the chickpea is also grown between 15 and 35°N latitude, and primarily between 20 and 30°N. Both chickpea and pigeonpea are mostly rain-fed crops. The pigeonpea, which is planted in the monsoon season, depends on the rain for its water requirements and is grown primarily as an intercrop. It often experiences drought stress during flowering and pod development because it matures during the dry season. In contrast, the chickpea is grown on the stored soil moisture received through rainfall during the monsoon season. Chickpea yields are strongly influenced by residual soil moisture from the preceding monsoon. The chickpea has four major phenological stages—germination, seeding growth, flowering, and pod development. There are different temperature optima for each. It, along with the pigeonpea, has indeterminate growth, with no distinctive vegetative or reproductive phases. Chickpeas are also a quantitive long-day plant. Long days promote flowering and short days delay it. Temperatures also influence flowering and pod set (Khanna-Chopra and Sinha, 1989).

Cowpeas

Whereas the production of chickpeas and pigeonpeas is primarily Asiatic and they flourish in the semiarid tropics, the cowpea is one of the most ancient of African foods, and is one of Africa's major food staples. Nigeria is the worlds' leading producer and the primary center for genetic diversity. Varieties have been developed with pods that are on top of the plant and more resistant to pests. A significant breakthrough at the International Institute for Tropical Agriculture (IITA) near Ibadan, Nigeria, is a determinate (sets pods uniformly) variety that matures in 60 days, and, like the chickpea in Asia, will grow on the residual moisture from a rice crop. Such early drought-resistant varieties are well suited to the hot-dry or humid conditions of the tropics, and require little water, weeding, or pesticides. They can be grown as a second crop where only one grew before. Cowpea production, however, is not limited to Africa. Extensive production areas are found in China and the southern United States. The genetic diversity of cowpeas affords considerable latitude in adaptation to many climatic regions in the tropics and semiarid and subtropics. They are cultivated as seed, vegetable, and fodder legumes from the semiarid to the lowland humid tropics. Cowpeas are grown in some very dry and agriculturally difficult places, and are seldom irrigated or fertilized with inorganic nutrients. They represent less than 2% of the total world production of grain legumes.

There are other legumes of significance as food crops. Garden peas and green beans, both pod and shell, are grown extensively. There are both determinate and indeterminate vine types. They may be found in many locations and climatic settings in Europe, Asia, Africa, North and South America, and Oceania. Depending on seasons, they may be grown in every state of the United States.

The dry field bean, of many genetic types, shapes, and colors, is an important food staple from Canada and in the northern to the southwestern United States, Mexico, Central America, and throughout South America, especially in Brazil. In North America, they are widely adapted from Ontario, Canada, Michigan, and New York to the southwestern United Sates. A major Collaborative Research Support Program (CRSP), involving research contributions from many states in the United States, along with many African and Central and South American countries has been in progress for over 14 years. The focus is on the specific agroecological regions of the world where bean and cowpeas are grown (The Bean/Cowpea Collaborative Research Support Program Executive Summaries of the Annual Reports, Michigan State University at East Lansing, 1992).

Broad beans, sometimes called the horse bean (*Vicia faba),* are grown throughout the European countries, especially those bordering in the Mediterranean and are an important winter crop in Egypt. The broad bean thrives in cool climates.

Among the important legumes are also the peanut and the closely related groundnut. They are grown throughout the tropics and subtropics and in the warmer of the temperate zones in mostly rain-fed areas. Peanuts originated in the tropics and are a thermophilic (responsive to high temperatures) crop. The heat sums or accumulated temperatures range from 2500 to 4800°C. China and India are the main peanut-producing countries of the world. India is number one in both production and in area planted. Lesser quantities are produced in the United States, Argentina, Senegal, and the Sudan. Peanuts in the United States and elsewhere are troubled by rusts, leaf spots, mottles, and wilts. All are climatically related. In China, the northern boundaries and climatic limitations of production are extended and maturity hastened, with substantial increases in yield, by plastic soil covers or mulches.

An important component of the rangelands, commonly referred to as grasslands, which comprise, along with adjoining semidesert, forest, and cropping areas, 47% of the worlds' land area, are the legumes. The rangelands are particularly important in China, Argentina, Brazil, the United States, Russia, and Australia. They are also prominent in the Middle East, many parts of Africa, North and South America, and Southeast Asia. Forty percent of China's land area is grassland. They provide a great variety of native herbages for cattle, sheep, goats, camels, horses, and many species of wild life. Many species of legumes are a part of rangeland ecosystems and some are very deep rooted and extremely drought resistant. The proportion of legumes in grasslands or pasture usually ranges from 5 to 10%. Legumes, commonly found in grasslands or pastures, are alfalfa, which is a perennial, and a wide variety of clovers, most of which are biennials, among which red clover usually predominates. Another important series of perennial legumes are the *Astragalus* species, which are among the most drought-, wind-, and blowing sand-resistant of all herbages. Still another important group of legume herbages, adapted for growth in temperate and cool climatic regions, are the lupines, extensively grown in some parts of Russia and the Ukraine.

There are many uncertainties as to the global effects of rising levels of atmospheric CO_2 on the productivity and yield of the above diverse kinds of seed legumes, and the legume components of grasslands. The experiments conducted thus far are extremely fragmentary and noninclusive (Paez et al., 1983). They are all C_3 plants, and if the soybean is a prototype for favorable results and is a representative model for response for other legumes, there should be significant increases in yield, root growth, total plant dry weights, water use efficiency, and biological nitrogen fixation. Indeed, Poorter (1993) declares, on average, the response of C_3 species capable of symbiosis with N_2-fixing organisms (both herbaceous and woody plants) is higher than that of other species. There is one complication with experiments conducted thus far. Many plant species, capable of symbiotic nitrogen fixation, will often not have had symbionts under the often nutrient-rich conditions used in these experiments.

If, however, the increasingly higher levels of atmospheric CO_2 are accompanied by significant changes in climate, as reflected by temperature and rainfall, the direct beneficial effects on photosynthesis, biological nitrogen fixation, and water use efficiency might be nullified, at least in part. Many legumes, such as the chickpea and pigeonpea, exhibit different temperature optima for various phenological stages (vegetative growth, flowering, fruit-set, seed development). A temperature-induced failure in fruit set might unduly extend the vegetative stage. Conversely, the seed legumes have shown wide climatic adaptability and have an even wider geographic distribution. This is reflected especially in field beans and cowpeas. It would seem that with the significance of seed legumes as global sources of vegetable proteins and edible oils for human and livestock production, both for food crops and components of rangelands, coupled with their capabilities for biological nitrogen fixation, possible increased root mycorrhization (Rogers et al., 1992a), and potential growth and yield enhancements thereof, by elevated levels of atmospheric CO_2, we should have some better inventory of the possible positive impacts on global food production.

Edible Oil Crops

Closely allied to some of the seed legumes, particularly the soybean and peanut, are the other edible oil crops—*cottonseed, rapeseed* (known as canola oil in Canada and the United States), *sunflower, and sesame.* Cotton, although not a major food crop, deserves special attention because of its global significance in world agriculture as a major fiber crop, its seed being a major source of oil and protein, the land, water, fertilizer, and pesticides devoted to its production, and because it is a major item, along with its products, in international trade. Many nations are financially dependent on it as an export item. Furthermore, as noted earlier, the crop responds in a remarkable manner to elevated levels of atmospheric CO_2. Herewith, many experiments, as reviewed earlier, have been, over the past 30 years, conducted on crops of cotton under controlled environments and in open field and free air conditions. These are still in progress (Kimball et al., 1993a).

There are important edible oil crops grown throughout the world other than those of leguminous origin. Rapeseed is among the oldest and most important. In China, records of its production go back almost

2000 years. Even today, China leads the world in rapeseed production, followed by the European community, India, and Canada. Both winter and spring types are produced in areas paralleling, in large part, the production of spring and winter wheat, with dates of seed sowing and harvest, comparable to spring wheat in the northern provinces and with winter wheat in the more central parts of China. Its large geographic distribution, extending from the tropics to the Arctic circle, is indicative of wide climatic adaptability and temperature responses. New first-generation hybrid varieties are much higher yielding and show extended climatic adaptation. With the crops' wide climatic adaptability, there is also flexibility in harvest times. Rapeseed (canola) in North America is successfully produced in Alaska, several provinces of Canada, and is an emerging potential oil crop in the United States, as well as in Eastern Europe.

The sunflower is also among the important edible oil crops of the world. Sunflowers originated in North America and were introduced to Europe and China about 400 years ago. Among the important sunflower-producing nations are Russia, Argentina, Eastern Europe, the United States, and China. The crop is widely adapted climatically. It is found in all the northern and central provinces of China, every state in the United States, including Alaska, and in the Ukraine, where very significant genetic improvements, including dwarf varieties and hybrids, have been developed. Other edible oil crops are sesame, flax, olive, oil palm, and the coconut. Sesame, being thermophilic, is produced only in tropical and subtropical climatic zones. India is the world's largest producer, with China second. Production in China is confined mostly to the central and southern provinces. Flax seed is also an important oil crop with a climatic requirement quite different than that of sesame. Of the approximate 4.7 million ha devoted to its production worldwide, the most important producing areas are in India, Russia, Canada, Argentina, and China. The olive is the major oil crop in Mediterranean countries, especially Spain, Italy, Turkey, and Israel. The olive tree is extremely long-lived and drought resistant and adapted to the semitropics. It is grown in many semidesert areas that do not have sufficient water for other crops. Two other important edible oil sources are the coconut and the African oil palm. Each is grown extensively in the lowland humid tropics of southeast Asia, South America, and Africa.

The effects of an elevated level of atmospheric CO_2 on the productivity of major edible oil crops, other than the soybean and some minor studies, including the sunflower (Carlson and Bazzaz, 1980; Hunt et al.,

1991; Mauney et al., 1978; Morison and Gifford, 1984) are not known. All, however, are C_3 plants indicating a positive response would be expected. What affects a higher CO_2 world would have on the oil content or the quality of the oil constituents are largely a mystery. The rising importance of the vegetable oils, being cholesterol free, in the human diet and their extensive production under widely varying climatic zones in both agriculturally developed and developing countries, should be a mandate for a research effort, as to both climatic vulnerabilities and to the physiological effects of the rising levels of atmospheric CO_2 of the global atmosphere. An additional incentive for action is that oil seed production, along with carbohydrates, is an end product of phyotosynthesis and the photosynthetic process is mediated by elevated levels of atmospheric CO_2.

Plantains and Bananas

Plantains and bananas are important food crops in both the lowlands and highland of sub-Saharan Africa. This is where 50% of the world's 68 million tons are produced (Buylsteke et al., 1993). Plantains are starchy bananas, which make up one quarter of the world production of bananas. Unlike the sweet dessert bananas, plantains are a staple food. They are fried, baked, boiled, pounded, and roasted, and are consumed alone or with other food. About 70 million people in West and Central Africa are estimated to derive more than one quarter of their food energy from plantains. The balance of the world's bananas and plantains is produced in South and Central America (35%) and Asia (15%) (Swennen, 1990). Southeast Asia is considered the center of origin.

Bananas are produced in most tropical and subtropical countries, but with a low temperature tolerance not below 10 to 12°C. Bananas and plantains require a hot and humid environment. Ideally, the average air temperature should be about 30°C and rainfall at least 100 mm/month. they are the most widely consumed fresh fruit in the United States and in many other countries. The response of this important tree food crop to elevated levels of atmospheric CO_2 is not known.

Another tropical tree food species is the *coconut.* It, along with the banana, is among the 25 most important world food crops. Likewise, with the banana, the response of the coconut to elevated levels of atmospheric CO_2 is not known, and, as with the banana, and to a lesser extent with rice, the climatic constraint is one of cold temperature, with no place on earth too hot or too humid.

The Sugar Crops

Sugarcane and sugar beet are among the 25 to 30 crops that stand between the world's people and starvation. They are vastly different in their optimal climatic requirements. Nevertheless, their simultaneous production has been observed at the same locations near the 33°N parallel in both India and China. Sugarcane, as a C_4 species, is acclaimed as a leading performer in the photosynthetic conversion of light to chemical energy. It is produced commercially from 36.7°N (Spain) to 31°S (South Africa), and from slightly below sea level to above 1000 m. A total area of 12 million hectares is devoted to its production in 79 countries and it is a major export item for many nations in Africa, Latin America, the Caribbean, and Asia. India has the largest area with three million hectares followed by Brazil and Cuba. The crop has a long growing season, a high water requirement, some salt and drought tolerance, but little cold resistance. It responds well to high fertility, irrigation, drainage, and abundant sunlight (Irvine, 1983). Of all environmental variables, temperature is probably the most determinate for production of sugarcane. Being the most tolerant of high temperatures of any of the major food crops, intact plants may survive at air temperature near 60°C. The record high temperature survival was in a commercial field in the Khuzistan Desert of Iran (32°N) with an absolute maximum of 52°C. In this area, the average maximum for July is 46.4°C and the average minimum is 32°C. Much of the world's sugarcane experiences temperatures of 26 to 30°C during the growing season. Cooler temperatures (26 to 30°C) slow stalk elongation and increase the sucrose content. Lower temperatures (18.3°C or less) inhibit flowering. No place on earth is too hot for sugarcane production if there is adequate moisture.

Sugarcane, as with corn, sorghum, and millet, is a C_4 crop. Unlike other C_4 food crops, however, the measure of productivity is sugar, a direct product of the photosynthetic process and photosynthetic water use efficiency. The expectation is that sugar production could be greatly enhanced at elevated levels of atmospheric CO_2, especially under hot, dry climates that frequently prevail where sugarcane is produced.

The sugar beet, a C_3 plant, has a decidedly different climatic requirement than sugarcane. Its production parallels that of the potato and is confined to temperate climatic zones in areas where a photoperiod of 13 to 17 h and temperatures of 15 to 18°C prevail during the growing season. There should also be an ample supply of water either through rainfall or irrigation. Major areas for production are in the northern hemisphere in western and eastern Europe, China, India, and North

America. Being a short season C_3 plant, its production potential per unit land area for sugar falls short of that for sugarcane.

The response of the sugar crops to elevated levels of atmospheric CO_2 is not well documented. There should be a strong positive response of the sugar beet, since it is a C_3 plant and the root being the marketable product, as in other root and tuber crops, should be an effective sink for photosynthetically fixed carbon. Experimental results confirm this conclusion (Ford and Thorne, 1967; Stulen and den Hertog, 1993). With sugarcane, the effect of added CO_2 was demonstrated by Hartt and Burr (1967), with a fourfold increase in net photosynthesis when the level in the atmosphere was increased from 100 to 600 ppm. Canopy levels of CO_2 may be markedly depressed at midday in sugarcane plantations when there is only light air movement above and still air beneath. This often occurs in many locations where sugarcane is produced. Since sucrose, whether in the beet or the cane, is a direct product of photosynthetic carbon fixation, the need for further studies on the effects of elevated atmospheric levels of CO_2 with these two major food crops should be apparent.

5.3 FOREST TREES AND FORESTS

Forests occupy one third of the earth's land surface. Their vegetation and soils contain about 60% of the total terrestrial carbon. They have historically contributed to human economic and social progress by providing shelter, fuel, food, watersheds, land stabilization, recreation, and a variety of commercial products (Clawson, 1979). Estimates are not available on the amount of land managed for agroforest systems. Globally, world forest systems are classified as temperate, boreal, and tropical (Wisniewski and Lugo, 1992).

A considerable number of reports have appeared concerning the response of specific forest tree species to both potential climate change and to elevated levels of atmospheric CO_2. Uncertainties exist, as they do with specific forest tree species, regarding both forest ecosystems responses to climate change and to elevated levels of atmospheric CO_2 (Evans et al., 1993; Hoffman et al., 1984; Jarvis et al., 1989; Sedjo and Solomon, 1989; Solomon and West, 1985). It has, for example, been maintained that forestry is more climate sensitive than agriculture because trees planted now may mature during the period of expected climate change (Williams, 1985). This assertion has been questioned by Reifsnyder (1989), in that the most weather-sensitive portion of a tree's life is its first year. Trees also may live for hundreds of years. This implies that during their lifetimes, they can and do withstand great climatic fluctuations. It has been further postulated that climate scenarios projected by different GCMs for a doubling of the atmospheric greenhouse gases would result in major shifts of the geographic ranges of such common North American trees as sugar maple, yellow birch, hemlock, and beech (Ruttan, 1990). Specifically, the northern limit would move north by 500 to 700 km, and the western range limits would retreat eastward. For example, the range of loblolly pine in the southern United States would shift north by several hundred miles (Smith and Tirpak, 1989a). Other species would decline or migrate (Urban and Shugart, 1989; Woodman and Furiness, 1989). Although there is some information available about the climate sensitivity of seedling tree types, little is known about stages of flowering, fruiting, and seed germination, which may also be climate sensitive.

The response of forest trees to continuously elevated levels of CO_2 and other greenhouse gases under open field conditions for whole forest ecosystems through long time periods for wet, moist, and dry conditions in boreal, temperate, and tropical life zones is likewise not known (Wisniewski and Lugo, 1992). The rising levels of atmospheric CO_2 do accelerate net primary productivity and have the potential for a net acceleration of ecosystem productivity. That this is, indeed, occurring is borne out by the progressively increasing amplitude of the annual carbon cycle recorded at the Mauna Lao Observatory in Hawaii. Forest tree responses to elevated levels of atmospheric CO_2 will likely determine the fate of many species (Graham et al., 1990).

Additional supporting evidence that the rising levels of atmospheric CO_2 is truly increasing the growing stock of forests comes from Finnish Forest Research Institute (Kauppi et al., 1992). They reported a 30% increase in forest growth in Finland, France, and Sweden between the early 1970s and the late 1980s, attributed not to reforestation but, in part, to a 9% increase in atmospheric CO_2 and with an estimated 25% larger growing stock in 1990 than in 1971 for all European countries. Results by Graumlich (1991) in the United States, however, were not as positive. A study of tree rings was made of three species, *Pinus balfouriana, P contorta,* and *Juniperus occidentalis,* with 20 to 25 trees from each of five sites, all at altitudes above 2700 m. The study extended back from the 1980s to between 1005 and 1570 A.D. Recent examples of growth stimulation were equalled or exceeded in the past when CO_2 concentrations were lower, with the conclusion that there is no strong evidence in support of the concept of enhanced tree growth by CO_2 fertilization since the onset of the industrial revolution. Growth factors other than that of atmospheric CO_2 could, however, have contributed to such differences.

Observations with loblolly pine and sweet gum species, along with corn and soybeans grown in the field in open-top chambers by Rogers et al. (1983b), revealed that increases in total biomass of tree species exposed to continuously elevated levels of CO_2 for 3 months ranging from 340 to 910 ppm, showed a ranking for tree species similar to corn and soybeans. The range was from 157 to 186% of control (340 ppm) values. Later studies by Luxmoore et al. (1986) established the following responses of Virginia pine to CO_2 enrichment during daylight hours only for 4 months under mineral nutrient-deficient soil conditions: greater growth, especially for roots; greater nutrient uptake associated with the greater root growth, especially of Ca and N and trace elements; greater nutrient use efficiency, particularly for P and K; and reduce P, K, and NO_3 concentration in the soil leachate.

As already indicated, many forest trees and shrubs, including all the legumes, have biological nitrogen-fixing potentials. These would include the many species of locust, the mesquite, and alder. Few observations thus far have been made on the responses of such trees to elevated levels of atmospheric CO_2. Growth responses appear to be positive or at the least not negative (Poorter, 1993).

As reported earlier, perhaps the most striking response of single species of trees to elevated levels of atmospheric CO_2 has been with *Citrus* and pine (Garcia et al., 1993, 1994; Idso et al., 1991; Idso and Kimball, 1992, 1993). The studies are still in progress. Trees are being exposed continuously to variously elevated levels of atmospheric CO_2 under field conditions in open-top chambers. Treatments are replicated and all are irrigated and fertilized to minimize any moisture or nutrient stress. Increases in growth, determined by trunk volume and resulting from increases in photosynthetic rates, have been phenomenal, ranging from 2 to 3.6 times the rates of trees grown under ambient CO_2 conditions.

Significant, also, are the results of studies by Norby and O'Neill (1989), Norby et al. (1992), and Gunderson et al. (1993) with yellow poplar (*Tiriodendron tulipifera)* and white oak (*Quircus alba)* conducted during 3 years of growth in open-top chambers, with elevated levels of CO_2 maintained both day and night. Both water use efficiency and net photosynthesis were significantly increased. There was no decrease in responsiveness to CO_2 enrichment over time. Photosynthetic enhancement was sustained with no differences in leaf senescence.

Fruit Trees

Many tree species, both in the tropics and temperate climate zones, are food crops, providing tropical fruits, nuts, and deciduous fruits in temperate zones of considerable economic importance and of increasing dietary significance. Among the most important, for the tropics, in addition to the banana, plantain, and coconut, which also harbor the wild relatives, are the orange, grapefruit, lime, lemon, avocado, guava, mango, papaya, and pineapple. Included also are Brazil nuts, cashews, macadamia nuts, palm oil, coffee, tea, cacao, cinnamon, clove, and vanilla. In the more temperate zones are found walnuts, filberts, pecans, and almonds and many tree fruits: apple, pear, plum, peach, apricot, and cherry as well as grapes and berries. World exports of coffee, citrus fruit, banana, palm oil, cocoa, and pineapple exceeded 20 billion in 1991 (Durning, 1989).

Ninety-five percent of the species of higher plants and all forest, fruit, and nut trees have C_3 photosynthetic metabolism, and most would be expected to respond dramatically to elevated levels of atmosphere CO_2, if experiments thus far conducted with citrus, pine, and several other species (Poorter, 1993) are any indication. These should include eucalyptus, radiata pine, Douglas fir, loblolly pine, black walnut, Scotch pine, aspen, and others that supply most of the world's wood. Cultivated fruit crops are also among the ones most adequately supplied with water and nutrients, are climate specific, and are protected from pests. Under such conditions, as proposed by some advocates, there should be few, if any, constraints to a positive CO_2 response. The evidence is also strong that continuous exposure of forest tree species and fruit trees to elevated levels of CO_2, simulating actual future anticipated conditions of a higher CO_2 world, would result in significant increases in productivity as reflected by the harvest index. The effects of extra atmospheric CO_2 on fruit quality are not known, nor susceptibility to insects, diseases, and nematodes. The likely result of a higher C/N ratio would be fruit of higher quality.

5.4 VEGETABLE CROPS

In contrast to fruit crops, both temperate zone and tropical, a considerable amount of greenhouse research has been conducted on some vegetable crops, particularly tomatoes, lettuce, and cucumbers, wherein the initial responses to elevated levels of atmospheric CO_2 for food crops were first recorded. Again, most leading vegetable crops—tomatoes, cucumbers, lettuce, cabbage, cauliflower, peppers, eggplant, melons, squash, okra, asparagus, green beans, and radish—have C_3 photosynthetic carbon metabolism, with the exception of sweet corn, one of the minor vegetable crops of the world. The expectation is that both photosynthetic activity and water use efficiency would be significantly increased. This has been confirmed in numerous greenhouse and field experiments.

5.5 RANGELANDS, GRAZING LANDS, WETLANDS

Other ecosystems of the biosphere under CO_2 enrichment may influence directly or indirectly the productivity of agricultural crops, livestock, and fish. The longest running field experiment to date with elevated atmospheric CO_2 in open-top chambers (Arp, 1991; Leadley and Drake, 1993) is being conducted on three plant communities in a brackish Chesapeake Bay wetland in the United States. Treatments began in the spring of 1987 and will continue through the 1994 growing season. The first 4 years of a long time study of CO_2 effects have now been summarized (Drake, 1992a, b, c). The response of two species has been followed: *Scirpus olneyi*, a C_3 sedge, a C_4 grass, *Spartina patens*, with C_4 photosynthetic metabolism, and a mixed community of these two species and the C_4 grass, *Distichlis spicata*. There were many interacting responses with *Scirpus*, but little response for *Spartina*. For *Scirpus*, there have been increases in quantum yield and photosynthetic capacity, sustained stimulation of photosynthesis, above ground production, reduced dark respiration, increased numbers of shoots, rhizomes, and roots, reduced nitrogen concentration of all tissues and increased nitrogen fixation, nitrogenous activity, ecosystem carbon accumulation, and methane production. There was no evidence that after 4 years of exposure photosynthesis in *Scirpus* had been down-regulated (Arp and Drake, 1991). The net ecosystem gas exchange increase per unit of ground area, in a plant community dominated by *Scirpus*, was 50%. In a mixture of all three species, *Spartina* and *Distichlis*, biomass of the C_3 component increased over 100% and this was accompanied by decreased biomass in the C_4 component of the community. Elevated CO_2 reduced water loss, increased water potential, and delayed senescence in all three species.

Drake (1992b) and associates (Arp, 1991; Arp and Drake, 1991; Arp et al., 1993) have come to other significant conclusions relative to plant responses to elevated atmospheric CO_2, namely the effects of temperature, dark respiration, and plant responses in field versus controlled environments. Reference is made to Baker et al. (1992a, b, c) and Allen et al. (1991b), wherein with rice, elevated CO_2 had the greatest relative or canopy photosynthesis at the highest growth temperatures (40 to 33°C). Also, the available evidence from numerous ecosystem studies and other observations suggest that plants growing in widely different temperature regimes should have different responses to CO_2. This, as mentioned earlier, may explain the relatively smaller or negligible responses of canopy photosynthesis to a doubling of the normal ambient CO_2 in a widely quoted study of the tussock tundra as reported by Oechel and Strain (1985) and Grulke et al. (1990). These plants, under arctic conditions, quickly acclimated by down-regulation of photosynthesis (Tissue and Oechel, 1987). In other studies, Overdieck et al. (1984) observed that in a mixture of perennial rye grass (*Lolium perenne)* and white clover (*Trifolium repens),* there was an increase of 25 to 45% in photosynthesis at 600 ppm of CO_2 over the long term, compared to 340 ppm. Additional studies by Nijs et al. (1988) revealed that with *Lolium perenne,* where CO_2 enrichment was provided throughout the growing season, both root growth and the yield of clippings were greatly enhanced.

It has been further suggested by Drake (1992a) that the down-regulation of photosynthesis in many plants, when exposed to elevated levels of atmospheric CO_2, may be, in fact, an artifact induced by restricted root growth and soil volumes in many controlled environments. Plants in the field are not so restricted, have greater soil volumes, and the greater root growth can accumulate additional carbon. Field studies, including his own, suggest the positive effects of elevated CO_2 in the field may be greater than has been estimated from plants grown in controlled environments. Recent experiments from a number of sources tend to support such a conclusion.

Drake (1992a), referring to basic plant physiological principles long enumerated by others, states that a fundamental property of green plants is that the rate of photosynthesis is dependent on the ambient CO_2 concentration. And there is overwhelming experimental evidence that this increases production in most C_3 plants. Hundreds of experiments with many species have confirmed this. Woodward (1993) supports this conclusion with the report that some plants such as *Boehmeria cylindrica* show a linear growth response to increasing concentrations of atmospheric CO_2. External environmental factors, such as temperature, light, and the availability of nutrients, may modify this response. Long (1991) points out that under warm conditions, there is little evidence of feedback inhibition being a major factor in plant response to elevated atmospheric CO_2. The greatest stimulation from high CO_2 on photosynthesis and growth can be expected to occur at high temperatures, with much lesser effects at low temperature (Table 4.1). Possible exceptions are rice and soybeans (Baker and Allen, 1993b). External factors that restrict growth, such as light, temperature, low moisture, and nutrients, will reduce, but usually not eliminate, the stimulation of production with CO_2, even when nitrogen is severely limited. There are also direct effects of the ambient CO_2 concentration on dark respiration with an immediate reduction in the rate of CO_2 efflux or O_2 consumption when the CO_2 around plant tissue is increased (Drake et al., 1994). *Although there have been very few long-term field studies of the effects on whole plants and ecosystem processes where the increased CO_2 was continuously maintained with suboptimal levels of other factors controlling growth, these data are consistent in showing an increase in plant production with an increase in CO_2 concentration of the ambient air.*

A large component of the earth's surface, a distinct ecosystem and of great importance as a food-producing system are grazing lands. They comprise 47% of the land surface. There are 3×10^6 ha of permanent grasslands—more than twice the area of arable cropland. These grasslands provide the food base for more than 1.3×10^9 grazing livestock. Their major role is a food-producing one, although they are important for water yield, recreation, wildlife, and genetic diversity. Worldwide, over 90% of the feed for food animals comes from forages. In the future, grazing lands will become more important as a source of food, either in the form of meat and milk, derived from livestock, or for the products of dryland agriculture (Pendleton and van Dyne, 1982).

The world's grasslands extend from the mid-latitudes of the United States (38°N), Russia (45°N), China (48°N), and South American pampas (38°S), to the tropical savannas and arid grasslands of the Middle East, Africa (12°N), and Australia (22°S), to the monsoon grasslands of India (26°N). They are found over a wide spread of temperatures and moisture. While C_3 grasses and legumes predominate, there are many grasses in the tropics and mid-latitude having C_4 photosynthetic metabolism (Pendleton and van Dyne, 1982).

Legumes, however, all having a C_3 photosynthetic carbon metabolism and with biological nitrogen-fixing capabilities, predominate among the world's forage crops, but not the world's grasslands. Alfalfa is the most widely produced forage crop, followed by clovers, lupines, and vetches. Though little research has been conducted on legume forages as to their response to elevated levels of atmospheric CO_2 (Cure and Acock, 1986; Krupa and Kickert, 1993), their responses have been positive as to photosynthetic efficiency, water use efficiency, yield increase, enhancement of root growth, and biological nitrogen fixation. Any growth and yield responses to CO_2 would not likely be limited by soil nitrogen shortages.

Data on the *rangeland/grazing land* responses to elevated atmospheric CO_2 are hard to come by. Only recently Owensby et al. (1989, 1993;

1994) reported that, in general, rangeland biomass increased when open-top field chambers were maintained continuously at or near double the atmospheric level of CO_2 during the entire grazing seasons from 1989 to 1991, located in a pristine tall grass prairie of both C_3 and C_4 grasses, north of and adjacent to Kansas State University, Manhattan, in the United States. There was a 30 to 40% increase in biomass for the predominant C_4 tall grass species, but not for the C_3 species. This may have been related to below normal precipitation. It also suggests that over time, community composition might shift. For the dominent tall grass C_4 range species *Andropogon gerardii,* there was a 50% reduction in stomatal conductance and a much more rapid response to variable sunlight. Knapp et al. (1994) reported that these direct effects on plant water relations would apply to all ecosystems. While the concentration of nitrogen in the forage grasses from the CO_2-enriched chambers was less, the total nitrogen absorbed was substantially greater. It is obvious that with the significant role that rangelands play in the total food chain of the earth, there are many unknowns concerning their response to both climate change and elevated levels of atmospheric CO_2. Neverthe-less, CO_2 enrichment reduced the effects of water stress on both C_3 and C_4 grasses (Sionit and Petterson, 1986), and above ground net primary production was increased on most sites in the United States with doubled CO_2 and three climate change scenarios (Baker et al., 1993).

5.6 AQUATIC PLANTS

Of the aquatic plants, whether of emergent foliage (obtaining most of their CO_2 from the atmosphere) or submerged (used dissolved CO_2 and HCO_3 ions), rice is by far the most important and best known food crop. Both its climatic adaptations and response to elevated levels of atmospheric CO_2 are reviewed elsewhere in this chapter. Other emer-gent foliage aquatic plants of much lesser importance as food crops include the cocoyam, consisting of taro *(Colocasia esculenta),* and tannis *(Xanthosoma sagittifolium).* Important for China (Wittwer et al., 1987) are lotus roots, arrowhead, the aquatic sweet potato, known also as swamp cabbage, the Chinese water chestnut *(Eleocharis tuberosa),* sometimes called "matai," and water bamboo or wild rice *(Zizania caduciflora).* Of the above crops, cocoyams (the edible species of *Colocasia* and *Xanthosoma)* are the most important in Oceania and in Africa, princi-pally Nigeria where 280,000 ha are produced of a world's total of over 1,000,000 ha. Cocoyams, along with the aquatic vegetables listed above for China, can be grown under saline water-logged conditions not conducive to the cultivation of most other crops in the tropics and subtropics, under a relatively wide range of temperature and soil environments (Nweke, 1987; Onwueme, 1987).

Responses of aquatic plants, especially the above food crops, to elevated levels of atmospheric CO_2 are less well known. Some results have been summarized (Wetzel and Grace, 1983; Wittwer, 1985). Few critical experiments have been made to determine the long-term effects of higher CO_2 on aquatic plants under field conditions. It is likely that emergent aquatic species will show the most pronounced positive effects (Wetzel and Grace, 1983). The studies of Drake (1992a, b, c) and Associates (Drake and Dahlman, 1994; Arp et al., 1993) on salt marsh wetland plant communities are an exception. Here sustained stimulation of photosynthesis and an increase in above ground production of 50 to 100% occurred over a period of 5 years with a C_3 sedge, *Scirpus olneyi*. Further details are given elsewhere in this chapter.

There are additional studies showing positive CO_2 responses to emergent aquatic species. Elevated levels of atmospheric CO_2 have remarkably sustained stimulating effects on the growth of both the water hyacinth (*Eichhornia crassipes)* and water lilies (*Nymphaea marliacea carnia)* with no appreciable change in plant dry matter content (Idso, 1989b; Idso et al., 1986, 1990), and the extra CO_2 appears to be most effective in enhancing plant growth if it is not proceeding at its maximum rate under ambient conditions (Idso, 1989b). Biomass in the water hyacinth under environmentally controlled conditions was increased by at least 40% at an atmospheric CO_2 level of 640 ppm. A 20-fold increase in net photosynthesis was reported upon raising the atmospheric CO_2 level to 800 ppm (Reddy and Smith, 1987). *Azolla-Anabaena,* a miniature free-floating aquatic fern grown for centuries in Northern Vietnam and Central and South China, is a nitrogen-fixing green manure crop for rice and a fodder for pigs. This aquatic species not only responds dramatically to a doubling of the ambient CO_2 atmospheric level, but greatly reduces the debilitating effects of temperatures above 30°C (Idso et al., 1986, 1990). Some truly remarkable responses were obtained by Idso et al., 1990) with hardy water lily (*Nymphaea)* cultivars grown outdoors in sunken tanks within open top chambers at 650 ppm of CO_2, compared with 350 ppm for a 5-month period. Differences in 25 plant characteristics were recorded among which were a 49% increase in net photosynthesis, 18% increase in leaf size, and a whole-plant biomass enhancement of 270%. The effects of even greater levels of CO_2 are not known. It may be speculated, however, that a doubling in atmospheric levels would increase the concentration of CO_2 and H_2CO_3 in the oceans and this might significantly increase the rates of photosynthesis (Holm-

Hansen, 1981). In summary, one must conclude that photosynthetically active aquatic plant species are among the most responsive of plants to elevated levels of CO_2.

5.7 OTHER CONSIDERATIONS

Concerning fruits and vegetables in temperate zone agriculture, it is reasonable to assume that a warmer climate will increase both the length and the intensity of the growing season, the growing degree units, and frost-free periods in both the northern and southern hemispheres. According to what now appear to be flawed models, however, the most pronounced effects would be near the poles and specifically the North Pole. For North America, the greatest impact should be in Canada and Alaska. Further, the risk of freezing temperatures in Texas, Florida, California, and Arizona for citrus, tomatoes, and other frost tender crops should be reduced (Wittwer, 1994).

There have been many additional speculative projections as to the effects of a presumed global warming on crop productivity, usually without factoring in the effects of an elevated atmospheric level of CO_2. For the United States, farmers in Minnesota might see yields of corn and soybeans doubled, while in states further south, production would be reduced. Michigan, New York, and Pennsylvania and southern Ontario would become even more important than they are now in fruit and vegetable production, with more serious consideration and perhaps need of irrigation. Crop production in the southern states could shift more to citrus and other tropical fruits and possibly more sugarcane. Yields of major food and feed grains in the great plains could drop with projected shortages of water for irrigation. Similar projections would hold within other major food-producing nations including China, India, Russia, Spain, Brazil, Argentina, Chile, Australia, and New Zealand. A Northern Hemisphere or southern migration in the Southern Hemisphere would likely increase the need for and use of irrigation and fertilizers on what would likely be sandy soils. This could create or worsen already existing groundwater problems. Higher crop yields would likely require more fertilizer and water. The two greatest global climatic constraints for food crop production on earth today are that, first, the continental climates are too cold, and second, there are inadequate supplies of water. A global warming would tend to alleviate the first but the effects on the second are not known.

Agricultural trade, especially with Russia and China, could be altered. Global warming would help Russia boost wheat production by perhaps 50% if a climate much like that of southern Canada were to occur. Crop production, likewise, in China should be greatly enhanced, providing additional water was available, with the northern migration of soybeans, winter wheat, rice, corn, and cotton. Finally, assuming the climate models have a degree of accuracy, there should be little change in the tropics. Thus the agricultural crop productivity impact on most developing countries in Central and South America, Africa, India, Southeast Asia, Indonesia, and the islands of the Pacific should be minimal. Some regions and crops will be more vulnerable than others. North America, with its natural resources, is strategically critical to the stability of world food supplies (Chameides et al., 1994). Adequate resources for crop and livestock production are usually most critical in agriculturally developing countries (Oram, 1985).

5.8 LIVESTOCK AND POULTRY—THE FOOD ANIMALS

Three fourths of the protein, one third of the energy, most of the calcium and phosphorus, and substantial amounts of essential vitamins and other minerals in the American diet are from animal products (Pond et al., 1980). The direct effect of the rising level of atmospheric CO_2 is innocuous as far as food animals are concerned, but directly affect the world's food supply by increasing the photosynthetic and water use efficiencies and yield of forages found in pastures, grasslands, ranges, and meadows, and for feed grains, which constitute most of their feed. Herein, there is almost a direct parallel related to food production since food animals primarily consume plants or plant products, which in turn, are converted to meat, milk, and eggs. On a world-need basis, 90% of the feed for livestock comes from forages (Pendleton and van Dyne, 1982). Similarly, the effects of a global warming or climate change, as it might indirectly impact crop productivity of the feed-producing areas of the world, would likewise indirectly impact animal productivity.

Two thirds of the world's agricultural land is for forage production. Sixty percent of this area is not suitable for growing cultivated crops. Ruminant animals, in particular, harvest this vast resource. They are natural protein factories, a living storehouse of mobile food. They constitute a global reserve food supply that exceeds that of grain.

A significant global climate change or warming would affect the production efficiency of both livestock and poultry and likely their global distribution. Climate is very important to all animal-production systems because it affects both the animal directly and influences the amount of forage and feed production and its distribution. A climate change caused by the continual rise in atmosphere CO_2 can produce changes in climate risks and, thus, induce changes in animal stress for any time or place. Moreover, climatic fluctuations are among the foremost in their effects on both forage quantity and quality. Drought can be catastrophic to animal production and survival. Rapid temperature changes increase animal stress. This is particularly true for animals in feedlots that are not properly designed for limiting animal stress during extreme conditions. Each year millions of kilograms of animal products are not achieved because of adverse weather.

Each species of livestock and poultry is impacted in its performance by the environment. Each has a comfort zone. Effective ambient temperatures below the critical temperature constitute cold stress and those above constitute heat stress. But any listing of critical temperatures for animals also requires specific description of animal variables. For example, a low critical temperature for sheep may change from 0°C in a sheep with fleece to 20°C for shorn sheep. Ames (1980) provides details as to critical low temperatures for cattle, sheet and swine, all subject to many variables. Temperature has pronounced effects on rates of gain, feed efficiency, disease incidence, and reproductive performances of all food animals. The importance of climate in livestock and poultry production, as a counterpart to crop production, is well recognized by producers. Controlled environments (windbreaks, confinement buildings—with or without heat) are often used, along with management options, by livestock and poultry producers to enhance production efficiency when they are exposed to stressful climatic environments. Hahn (1976), however, has emphasized that optimum environments for maximum production and feed efficiency are not necessarily optimal for economics and energy use. The production of dairy cows, finishing hogs, laying hens, and broilers, as a function of climatic factors, has been combined with the probability of occurrence of related weather events to predict production losses.

Climate has a major impact on determining geographic locations for successful livestock and poultry production. Dairy production is primarily a temperate zone type of agriculture with concentration in the United States and northern Europe. Meanwhile, however, it has become important in India, particularly in the Gujarat state with the water buffalo and its adaptation to the tropics (National Research Council, 1981), and the development of productive C_4 tropical grasses as a source of forage, coupled with farmers' cooperatives and a ready market. Beef cattle, likewise, thrive best in temperate climatic zones, but they are found in many tropical and subtropical areas. Laying hens, broilers, and turkeys are commercially produced from Texas to Minnesota and various species of laying hens and ducks are found throughout China. Goats and camels are adapted to the warm and arid climates of the Middle East, China, and India. Losses from poultry and in milk production can be extensive during heat waves and other extreme events of weather. In general, livestock and poultry production can respond to climate change with close linkage between crops and animals. This is especially important in the developing world. Livestock and poultry buffer the impact of changes in weather, with expansions in good times when feed is adequate, and culling back in bad times (Waggoner, 1992).

Perhaps the greatest of all global impacts of climate on food animals relates to animal diseases and vectors of animal diseases, many of which may result in greater prevalence and migration associated with any projected global warming. The possible increased incidence of animal diseases as a result of a significant global warming has implications for the United States and world populations. Stem et al. (1989) reviewed in detail animal disease patterns induced by a possible greenhouse effect for blue-tongue, anaplasmosis, Rift Valley fever, and African swine fever. Such diseases and others are currently kept in check largely because of restrictions, climate places and vectors, environmental habitats, and the disease-causing agents themselves. Similarly, climatic warming under a rising atmospheric CO_2 level would extend both the time of activity and intensity of the horn fly on pastured cattle in the United States. Activity would extend both earlier in the spring and later in the fall. The result would be losses in weight gain in beef cattle, and reduced milk production in dairy cattle (Schmidtmann and Miller, 1989). Outbreaks of foul cholera in turkeys has been linked to climate conditions (Hahn, 1976). Any global warming as related to the prevalence and severity of diseases would likely be deleterious rather than beneficial.

5.9 GLOBAL IMPLICATIONS FOR DEVELOPED AND DEVELOPING ECONOMIES

Irrespective of technological advances in agricultural productivity, weather and climate on a global perspective still are the most determinant factors in agricultural productivity.

The most comprehensive studies of climate and its impacts on global agricultural productivity and food security are those of Parry and associates (Parry, 1990; Parry et al., 1988a, b; Rosenzweig and Hillel, 1993b; Rosenzweig et al., 1993b; Rosenzweig and Parry, 1994). Most of the recent research on climate and food has focused on regional and national assessments of the potential effects of climate change resulting from an anticipated greenhouse effect from a doubling of the atmospheric CO_2 projected to occur about the middle of the 21st century (Allen and Gichuki, 1989; Mearns et al., 1992; Peart et al., 1989; Ritchi et al., 1989). These studies have been in more or less isolation without consideration of production in other places, and patterns of trade among nations, agricultural development, and food security. Sensitivity studies of world agriculture to climate change (the impacts on agriculture are considered to be among the most striking) suggest that the effects of moderate changes on local and global economies may be small because losses of production in some areas would likely be balanced by gains in others (Cooper, 1982; Kane et al., 1992). Climate change does not appear to threaten global food security, but could be a serious risk for specific regions or countries. Only recently have the direct effects of elevated levels of atmospheric CO_2, as an enhancement of photosynthetic CO_2 fixation and water use efficiency, been factored into important studies, this is not withstanding the lack of an integrated, combined biophysical and economic assessment of the potential effects of climate change, coupled with the direct effects on plant growth of a changing atmosphere involving unprecedented increases in CO_2 and other greenhouse gases. Moreover, inputs of new technologies, which will surely occur, and the ability of farmers to adapt have not been on the agenda.

In one of the latest and most comprehensive studies (Rosenzweig and Parry, 1994; Rosenzweig et al., 1993), scientists from 18 countries (Argentina, Australia, Bangladesh, Brazil, Canada, China, Egypt, France, India, Japan, Mexico, Pakistan, Philippines, Thailand, Uruguay, United States, Russia, and Zimbabwe) used crop models and three climate change scenarios from three GMCs: the Goddard Institute for Space Studies (GISS), the Geophysical Fluid Dynamics Laboratory (GFDL) of the United States, and the United Kingdom Meteorological Office (UKMO) models, projected to the year 2060, when atmospheric CO_2 is presumed to double over current levels. The crops modeled were wheat, rice, maize, and soybeans. The crop models were run for current climatic conditions, for arbitrary changes in climate (+2 and +4°C) increases in temperature and ±20% precipitation, and for climate conditions predicted by the GCMs for a doubled atmospheric CO_2 level. The photosynthetic ratios (555 ppm CO_2/330 ppm CO_2) for soybean, wheat and rice, and maize were 1.21, 1.17, and 1.06, respectively. Changes in stomatal resistance (sm) were set at 49.7/34.4 for C_3 crops and 87.4/55.8 for C_4 crops.

Results of this study, with the direct effects of CO_2, and precipitation held at current levels, revealed that average crop yields, weighted by natural production, showed a positive response to +2°C warming, with yields of wheat and soybeans increased by 10 to 15% and rice and maize by 8%. Yields of all four crops turned negative at +4°C warming. This indicated a threshold of compensation for the direct effects of CO_2 between 2 and 4°C. Rice and soybeans were the most negatively affected. Results of this study showed a decrease in food security in developing countries. The worst scenario depicted is of severe climate change, low economic growth, and little farm adaptability. The agricultural sector should be encouraged to develop crop breeding and management programs for heat and drought resistance.

In summary, and for consideration of limitations relative to the results of these studies, are the following: The climate change scenarios were at the upper end of GMC projected warmings, with warming of 2 and 4°C projected to be greater at high latitudes in winter, rather than global means throughout the year (IPCC, 1990). Climate change induced by increasing greenhouse gases is likely to affect crop yields differently from region to region across the globe. The favorable effects on crop yields in temperate regions will depend, to a large extent, on full recognition of the potential benefits of elevated levels of CO_2 on crop growth. Significant areas of China, India, Indonesia, and Vietnam, with near one third of the world's populations and over one half the developing world, were not included. Many crops as staples for developing countries were omitted—including all the roots and tuber crops. There were no legumes except soybeans, and only some of the major oil crops, and the tropical and sugar crops (sugarcane and sugar beets) were not on the list.

Further studies by Downing (1992) were directed to climate change in vulnerable places and food security. Countries studied were Zimbabwe, Kenya, Senegal, and Chile. The studies were conducted without due consideration of the direct effects of CO_2 enrichment on global crop productivity, although a global average of 25% increase for C_3 crops and a 7% increase for C_4 crops was projected (Rosenzweig et al., 1993) and an assumption of less precipitation. It was suggested that for Zimbabwe with a 2°C increase in temperature, the core agricultural zone would be decreased by one third and further marginalization would increase the risks of crop failure. For Kenya, the projected food production for higher elevations would increase with a warmer climate, but the effects in the semiarid and subhumid areas could be devastating.

In Senegal, with climate change, an additional one million people may not be supported by rain-fed agriculture and the prospects for agricultural development would be exacerbated; and for Chile, a warmer environment would entail increased irrigation and possible dramatic shifts in river basin hydrology.

Not considered on a global scale thus far have been other possible benefits of rising levels of atmospheric CO_2. These include improved productivity of most food legumes, oil crops, all root and tuber crops, sugar crops, and tropical fruits and nuts, including the banana, plantain, and coconuts (Table 5.1). All are staple foods for hundreds of millions of people in agriculturally developing countries. To the above beneficial effects on crop yields can be added, in many instances, an enhancement of biological nitrogen fixation and the often compensating effects of higher levels of atmospheric CO_2 on low light deficiencies, both high and low temperatures, moisture deficiencies and nutrient inadequacies, salinity, and air pollutants. These limitations in food production are equally as important, if not more so, for the agriculture of the developing world as in developed nations. Contrary to published expressions of others (Bazzaz and Fajer, 1992; Rosenzweig and Hillel, 1993b; Strain, 1992), it is not likely from the above evidence and observations that the impacts of a global warming induced by rising levels of atmospheric CO_2 will be any more devastating on the developing world than in developed economies. There is no reason to believe that the developing countries, as a group, will be exposed to more climate changes than the developed countries. However, the margins of survival may be smaller, and the impacts more immediate (American Society of Agronomy, 1990; National Academy of Sciences, 1992).

6

Other Implications of Climate Change and Food Security

6.1 STRATOSPHERIC OZONE DEPLETION

Equal to the intensity of the debate over global warming and the so-called "greenhouse effect" is that of the growing ozone hole over Antarctica and the presumed deleterious effects of stratospheric ozone (O_3) depletion on plant growth, animal behavior, worldwide agricultural productivity, all natural and aquatic and terrestrial ecosystems, and human health hazards resulting from increased ultraviolet-B (UV-B) radiation (280 to 325 nm) at ground level.

All of this reached a political climax February 3, 1992, when NASA held a news conference announcing that the probability of significant ozone loss taking place in any given year is higher than believed before. This touched off a speech by then Senator Albert Gore before the Senate, raising the specter of an ozone hole over Kennebunkport, President Bush's summer home in Maine, followed by editorials in the *New York Times* and *Washington Post* and a 96 to 0 vote in the Senate calling for a dramatic stepping up of action for abatement (Morrison, 1992). A month later a senior NASA official announced, "There is no ozone hole over Kennebunkport. There never has been an ozone hole over Kennebunkport, and I don't really expect one."

Conversely, tropospheric O_3 is currently and has for over two decades been recognized as the single most ubiquitous phytotoxic regional air pollutant worldwide (see Chapter 4) for many crops (Manning, 1990) and forest trees (Adams et al., 1984; Adams and Taylor, 1990), and may already be responsible for a 15% decrease in the production of some of them, including, in particular, the seed legumes (soybeans and other legumes), some C_4 and C_3 grain crops, the potato, cotton, and pine trees (Heck et al., 1982, 1988; Kickert and Krupa, 1990; Krupa and Manning, 1988; Krupa and Kickert, 1993; Krupa and Nosal, 1984). Furthermore, it has been estimated that 10 to 35% of the world's grain production may occur in regions of the northern hemisphere whose ozone pollution is now most serious and may triple by the year 2025 (Chameides et al., 1994). Thus agricultural food production faces a double dilemma—too much ozone on the one hand and too little on the other.

Of prime concern is a review of first, the separate effects of UV-B radiation and ozone on world food crops and other vegetation. Second, are the interactions or joint and simultaneous effects of the rising level of atmospheric CO_2 and the effects of stratospheric ozone depletion, coupled with the accompanying atmospheric pollution of ozone. Finally, are the joint affects of altered ambient atmospheric CO_2, UV-B radiation, and O_3, and temperature and moisture levels on current and future plant growth, and worldwide agriculture, forestry, and other ecosystems productivity. The latter circumstance is reflective of current happenings in the real world and major issues relating to food security, of the total biological productivity component of earth, and the health and welfare of the human race (Fiscus et al., 1995; Krupa and Kickert, 1993).

Both CO_2 and UV-B radiation are expected to increase simultaneously with future changes in global climate. The atmospheric CO_2 level is projected to double from its current 360 ppm by the middle of the 21st century, and the rise of chlorofluorocarbons (CFCs), methane (CH_4), and nitrous oxide (N_2O) may substantially deplete the stratospheric ozone, which controls the amount of ultraviolet-B radiation (UV-B, between 290 and 325 nm) reaching the earth's surface. Hence, it is extremely important to quantify the photosynthetic and growth responses and yield outputs from simultaneous increases in CO_2 and UV-B radiation, especially among major food crops. Fortunately, some studies examining these simultaneous changes in the atmospheric environment have recently been completed, but with no unanimity of conclusions.

Coupled with all the above is the generous visibility provided by the press and large segments of the scientific community during the past decade to global ozone depletion, and conclusive linking of this decrease to a rise in atmospheric chlorine from CFCs. They, along with CO_2, CH_4, and N_2O, are the greenhouse gases that are projected to cause the widely publicized global warming. The CFCs and N_2O are reportedly depleting the earth's protective stratospheric ozone layer designed to reduce the harmful ultraviolet rays emitted from the sun and falling on the earth's surface. Their long life-times permit photo disassociation and the release of NOCL, which catalytically destroys ozone (Teramura,

1990). Ozone depletion and the growth of the Antarctica ozone hole reached a new low in 1993 (Kerr, 1993b). This depletion is now being credited with a whole array of detrimental physiological and biochemical effects on plants, food animals, and the human race. Of special interest are the effects on the major food crops (cereals, legumes, root crops) that feed the world and the major forest tree species. Our concern in this book is directed toward an assessment of the effects on plant growth, crop productivity, and all food-producing systems.

Most of the earth's land area with food-producing capacity and the forests are in the Northern Hemisphere and the tropics. Kerr and McElroy (1993) indicated that the first conclusive evidence for a downward trend in stratospheric O_3 levels was reported in 1985 where springtime values over the Antarctica had declined by 40% between 1975 and 1984. Negative trends in O_3 levels at other locations have since been noted. Kerr and McElroy (1993) further report that the detection of a long-term change of UV-B radiation at the earth's surface is much more difficult than measuring long-term declines in O_3. *The intensity of UV-B radiation depends not only on O_3 levels, but clouds, aerosols, haze, pollution, and ground albedo.* This is relevant to some of the conflicting results on plant growth to be reported. Stolarski et al. (1992) recently reported the statistically significant downward trend in atmospheric O_3 over much of the Northern Hemisphere. An approximate 1.0 to 1.4% decrease per decade has now been recorded for most of the crop-growing season at mid-latitudes. The depletion is predicted to continue at a rate of about 0.4% yearly. The tropics, where the ozone layer is naturally the thinnest, receive the highest UV-B levels. It is in the tropics where most of the rice and root crops, all coconuts and bananas, and much of the millet, sorghum, sugarcane, and seed legumes are also grown, and where the tropical rain forests are located along with at least half of the world's population.

A flood of reports beginning with the early 1980s on the deleterious effects of increased exposure of plants to UV-B radiation has appeared. Most of our knowledge of the effect of UV-B radiation has thus far been derived from agricultural crops and a few forest tree species. The visible and external impacts on plants are indicated to be reduced growth, smaller leaves, less biomass, thickened leaves, shortened internods, and smaller plants. For some economically important crop varieties, it is reported that food yields may be reduced by up to 25% for exposure simulating a 25% total column O_3 depletion. With cassava, for example, there was no effect on total biomass, but an increase in shoot/root ratios (Ziska et al., 1993). This could be significant for cassava where yields of roots are important and constitute the harvest index. It has been emphasized that such reductions would significantly affect people in areas, mainly developing countries in the tropics, where food shortages already occur (Krizek, 1975; Teramura, 1983, 1987, 1990; Teramur and Sullivan, 1987, 1988; Teramura et al., 1980, 1990; Sullivan and Teramura, 1988, 1990, 1992, 1995; Sullivan et al., 1994). Additionally, Teramura and Sullivan and associates have listed an array of physiological and biochemical effects, including direct effects on the primary reaction of photosystem II, dark respiration, stomatal conductance, and the proteins, lipid, and carbohydrate composition of plant tissue. With intense solar UV there may be bronzing and chlorosis on the leaves of sensitive plants. Marked differences among crop species have been noted, and even within varieties of the same crop.

Little information exists for long-lived perennials, such as trees, and for noncultivated natural plant communities. Plants adapted to growth at higher elevations showed less injury, implying the presence of natural adaptation to UV-B. Three years of supplemental UV-B irradiation on seven seed sources for loblolly pine, simulating those that would be anticipated with stratospheric O_3 reduction of 16 and 25% reduced plant biomass by 12 to 20% at the highest simulated O_3 depletion level (Sullivan and Teramura, 1992). Further studies of Sullivan and Teramura (1994) on loblolly pine were conducted where depletions of stratospheric ozone were combined with elevated levels of atmospheric CO_2. Neither photosynthetic capacity nor quantum efficiency was altered by the elevation of CO_2. The UV-B irradiance reduced total biomass by almost 12% at both 350 and 650 ppm of CO_2. Biomass partitioning, however, was altered. With UV-B radiation at 350 ppm CO_2, dry matter was preferentially allocated toward shoots and toward roots at 650 ppm CO_2. Additional studies by Sullivan (1994) on the tree species *Pinus taeda* (loblolly pine) and *Liquidambar styraciflua* (sweet gum) revealed that the effects of UV-B radiation on photosynthesis on loblolly pine appeared transient, and with sweet gum, there was an enhancement of the rate of photosynthetic capacity in leaves, with a minimal effect on biomass. Clearly more work needs to be done.

As already indicated, the so-called "greenhouse" or global warming, arising from increasing concentrations of tropospheric trace gases, is of much national and international concern. Among the climatic variables of the atmosphere, elevated levels of CO_2, UV-B, and O_3 are known to have numerous direct effects on agricultural crop production. Other accompanying stresses could be a global rise in atmospheric and soil temperatures and increases in soil salinity. For the real world, knowledge of the joint effects of these variables on vegetation is important. Krupa and Kickert (1989, 1993) and Rozema (1993) have provided summaries of some effects on several food crops, including sorghum, soybean, pea, bean, potato, corn, wheat, rice, oats, lettuce, cucumber, tomato, and a number of other vegetable and flower crops, many of which are among the species most sensitive to the joint effects of these three variables.

Two very recent studies have examined the interactions (Sullivan and Teramura, 1994; Teramura et al., 1990) and combined effects (Van de Staaij et al., 1993) on specific crops. Teramura et al. (1990) conducted an examination of the effects on the photosynthetic characteristics of wheat, rice, and soybean. Carbon dioxide concentrations of 350 (ambient) and 650 ppm were maintained throughout the growing season in College Park, Maryland. They were initiated May 10, along with an increase in UV-B radiation, corresponding to a 10% O_3 depletion at the equator. Seed yields and total biomass increased significantly with elevated CO_2 in all three species, when compared to the control. With concurrent increases in UV-B and CO_2, no increase in either seed yield (wheat and rice) or total biomass (rice) was observed. Contrastingly, CO_2 increased seed yield and total plant biomass in soybean within the elevated CO_2, UV-B environment. A significant increase in photosynthesis, apparent quantum efficiency, and water use efficiency with elevated CO_2 was maintained in all three species. Combining an elevated UV-B radiation with high CO_2 eliminated the effect of high CO_2 on photosynthesis and water use efficiency in rice, and the increase in apparent quantum efficiency associated with high CO_2 in all three species. In summary, increased UV-B radiation may modify or negate high CO_2-induced increases in biomass, seed yield, and photosynthetic parameters. Van de Staaij et al. (1993), working with a single species, *Elymus athericus,* a C_3 grass, concluded that elevated CO_2 concentrations stimulated plant growth and biomass production by 67%, whereas elevated UV-B radiation had a negative effect on growth, reducing biomass production by 31%. Enhanced CO_2 combined with elevated UV-B radiation levels caused a biomass depression of 8% when compared with controls, and the UV-B-induced growth depression can be modified by a growth stimulus induced by high CO_2. Finally, the sensitivity to concurrent increases in CO_2 and UV-B radiation may be highly dependent upon species and cultivars and modified by microclimate. This has been dramatically illustrated with rice cultivars at the International Rice Research Institute (IRRI) in the Philippines (IRRI, 1993). Such sensitivities suggest that crops and their weedy competitors might also respond differently to concurrent changes in CO_2 and UV-B radiation. Since UV-B may modify CO_2-induced increases in photosynthesis and yield of the major food crops and the growth of forests, not accounting for the simultaneous increases in UV-B radiation could result in overestimates of photosynthesis and plant production based on the response to the rising level of CO_2 alone (Teramura, 1990).

The discussion of UV-B radiations on crops would not be complete without reference to the recent and perceptive observations and conclusions of Fiscus et al. (1994) and Miller et al. (1994) presented at a NATO workshop in Gainesville, Florida, June 1993. The focus of these reports was on the effect of UV-B radiation on soybean production, one of the reportedly most sensitive of all crops to UV-B radiation, and a major food crop, and already the object of many studies with elevated levels of CO_2, tropospheric O_3, and UV-B radiation. The two stresses of UV-B radiation and tropospheric O_3 on crops may cooccur. Most reports, heretofore, show a high sensitivity of the soybean to both O_3 and UV-B radiation with reductions in photosynthetic capacity, plant dry weight, leaf area, number of pods, and seed yield. Fiscus et al. (1995) and Miller et al. (1994) however, found no UV-B suppression of growth or yield or any measurable effects on photosynthetic gas exchange or electron transport characteristics within the three diverse soybean varieties observed. Treatment levels of biologically effective UV-B radiation simulated the increase in ground level UV-B for stratospheric ozone depletion up to 13%, equivalent to about a doubling of ambient UV-B. Ozone levels were well above the ambient.

Ultraviolet-B radiation did not affect soybean seed yield in any of the three varieties. No $O_3 \times$ UV-B interactions were noted for any yield component. Net carbon exchange rate, stomatal conductance, and transpiration of soybean leaves were not suppressed. On the other hand, O_3 treatments consistently induced visible injury, suppressed net carbon exchange and water use efficiency, accelerated reproductive development, and suppressed growth and yield. *They concluded that tropospheric O_3 poses a greater threat to soybean production than any projected levels of UV-B radiation. They further concluded that any projected changes in ground level UV-B radiation that might result from stratospheric O_3 depletion resulting in impairment of growth of soybeans were irrelevant to anything outside laboratory situations.* Their reviews of literature revealed that UV-B doses used in other laboratories are frequently underestimated, along with cloud cover, atmospheric aerosols, and seasonal differences, which further contribute to the under estimation of actual doses. *They conclude that increases in ground level UV-B, well in excess of current projections for* *the next century, will not constitute any direct hazard to soybean production.*

It seems that as experiments continue in time and intensity that the earlier reported harmful effects of UV-B radiation on plants resulting from a depletion of stratospheric O_3 become more uncertain. This is especially true with extended exposure of tree species (Sullivan, 1994; Sullivan and Teramura, 1994; Sullivan et al., 1994).

6.2 BIODIVERSITY

Related to global warming and the rising atmospheric levels of CO_2 and other radiative greenhouse gases is the issue of biodiversity. It represents an example of nonmarketed resources under threat of serious degradation. Many scenarios project a global warming that would be very rapid compared with prehistoric changes of similar magnitude, and result in the demise of many plant and animal species that are unable to migrate or adapt to abrupt climate changes. It is not only the importance of temperature increases alone to biological communities but the prospect of widespread changes in precipitation patterns. For many species, a change in water availability would have greater impact than temperature changes of the order predicted. A rise in sea level resulting from thermal expansion of sea water and a melting of glaciers and polar ice caps, though highly speculative, has also been projected as disastrous to many biota. A warming trend may also alter the ocean's vertical circulation, causing changes in up-welling patterns that sustain many marine communities, a trend that would be another possible hazard (Peters and Darling, 1985).

Attention has been directed to some sensitive forest tree species and herbivores that could not survive under the projected dietary and climate changes. The projected rate of change in climate associated with the greenhouse effect, according to some scenarios, could perhaps exceed the rate at which some natural forests could evolve and migrate (U.S. Department of Energy, 1989). Changes in temperature, precipitation, and atmospheric concentration of CO_2 may have direct physiological effects on the growth of forest trees. For an individual tree, the impact of climate change includes the direct effect of CO_2 on photosynthesis and water use efficiency. Although the longer growing seasons of a warmer climate would raise productivity, ill-adapted trees could suffer frost damage during their prolonged growth, and others might not be chilled enough to germinate. The sensitivity of individual trees to moisture is vividly demonstrated by the difference in species from the north to south slopes of many mountain ridges and from the dry, rocky summits to the wet marshes below (National Academy of Sciences, 1992).

Conservationists have argued that a loss of species, generously exaggerated by Myers (1983), combined with subtantial species extinctions in tropical habitats, temperate regions, and the tundra in the arctic, will diminish the gene pool available for developing new foods, medicines, and industrial products (Reid, 1994).

Historically, nearly all medicines came from plants and animals, and even today they remain vital. Traditional medicine forms the basis of primary health care for 80% of the people in developing countries or for more than half the world's population. Materials from more that 5100 species are used to prepare over 11,000 folk prescriptions in historical Chinese traditional medicine alone (Wittwer et al., 1987). Nearly 2000 plant species were used in the former Soviet Union for medicinal purposes (World Resources Institute, 1992–93). Currently, some 25% of prescriptions in the United States are filled with drugs where active ingredients are extracted or derived from plants. Sales of these plant-based drugs amounted to 4.5 billion in 1986 and an estimated $15.5 billion in 1990 (Reid, 1994). The importance of biodiversity for agricultural production was emphasized in the classical National Research Council Report, "Genetic Vulnerability of Major Crops" (National Research Council, 1972).

In conclusion, the impacts on biodiversity have received little, if any, attention in any of the major studies, workshops, symposia, or conferences heretofore held on climate change, the global warming issue, or the rising levels of atmospheric CO_2 and other greenhouse gases. There has been the suggestion, however, that species-rich ecosystems consume CO_2 at a faster rate than less diverse ecosystems, which suggests that loss of biodiversity may promote CO_2 buildup and speed global warming (Anonymous, 1993a). Countering this is the fact that in the natural landscape many species take in carbon and make food by photosynthesis. The loss of a few species would probably cause little change in the photosynthetic rate. Diseases removed first the chestnut and then the elm from eastern United States forests, but their photosynthesis was quickly replaced by other species. Climate change may eliminate some species from the natural landscape, but its diversity will protect such functions as photosynthesis (National Academy of Sciences, 1992), and future climate change may increase the rate of loss of biodiversity and increase the value of genetic resources. Pimm and Sugden (1994) draw these succinct conclusions: Complex feedback makes linking well-documented changes in atmospheric chemistry (increasing levels of CO_2) to the physics of global warming difficult. Linking either to the biology of diversity is even a greater challenge. No two plant species need respond the same way to laboratory-controlled increases in temperature and CO_2. Mixtures of plants respond even more differently.

6.3 RISE IN SEA LEVELS

Typical of the many declarations concerning the rising level of atmospheric CO_2 and the accompanying greenhouse effect or global warming are statements such as this: "This greenhouse effect may, by early next century, have increased average global temperatures enough to shift agricultural production areas, raise sea levels to flood coastal cities, and disrupt national economies...A rise in sea level in the upper part of this range (25–140 cm) would inundate low lying coastal cities and agricultural areas, and many countries could expect their economic, social and political structure to be severely disrupted" (World Commission on Environment and Development, 1987).

Some have suggested that projected global warming could lead to a disintegration of the west Antarctic ice sheet, most of which is grounded below sea level. If the climate warms and the warmer ocean water penetrates under the ice sheet, the release of ice from the sheet would accelerate. *It is concluded, however, that the melting of the ice sheet is very unlikely and virtually impossible by the year 2100. Estimates based on the combined oceanic and atmospheric GCM suggest that several hundred years will be required to affect this degree of warming.* It is concluded that any rise in sea levels during the foreseeable future will be a result of thermal expansion (National Academy of Sciences, 1992). As for the United States, a rise in sea level is not so much of an agricultural concern as it is for municipal water supplies in such low lying areas as New Orleans and New York City (Waggoner, 1990). In fact, a rise in sea levels was not worthy of mention in the task force report of 1992, "Preparing U.S. Agriculture for Global Climate Change" (Council for Agricultural Science and Technology, 1992).

This brings us to the several reports of Titus of the U.S. Environmental Protection Agency (Titus, 1990a, b; Titus et al., 1993). He concludes that the major impacts of sea level rise would affect agriculture, particularly in deltaic nations, such as Bangladesh, where a substantial part of the agricultural lands could be inundated or eroded. Increased flooding could threaten crops in many areas, and salt water intrusion would threaten irrigation water in many other areas. Although important for localized regions, it would be relatively insignificant on a worldwide basis compared with other projected impacts of global warming on agricultural production. Nevertheless, particular aquatic crops that are typically grown in low areas, such as rice and taro, may be disproportionately affected. Details of the implications of a possible sea level rise for the world's agriculture have yet to be delineated (Titus, 1990a).

6.4 ADAPTATION OF FOOD PRODUCTION TO CLIMATE CHANGE AS REFLECTED BY GEOGRAPHIC SHIFTS IN CROPS AND LIVESTOCK

Historically, many crops have been successfully introduced or have migrated into climate zones for which they were not indigenous or previously adapted (McNeill, 1991). Changes in average temperature are probably not as important for agriculture as changes in precipitation and evaporation. Farming has always been sensitive to weather. Farmers can adapt quickly to any likely rate of greenhouse warming. Countries such as the United States, Russia, China, Australia, Argentina, Brazil, Chile, and India have many climate zones and active and aggressive agricultural research programs. They could adapt quickly to any greenhouse warming. Poorer countries with fewer climate zones may have some difficulty (National Academy of Sciences, 1991).

For the United States, the economic effects of the MINK study (Rosenberg, 1992a, 1994) were calculated on two unlikely assumptions (National Academy of Sciences, 1992). First, that the climate would change immediately in the world as it is today; and second, that no adjustments to these changes are attempted. Simple adaptations to a shortened crop-growing season, imposed by a warmer world, would merely be to use later maturing varieties and earlier plantings. With a selection of the most appropriate planting dates and cultivar for the changed climate, yield reductions could be minimized, and this without considering the direct positive effects of CO_2 fertilization, where the outlook would be even more optimistic. Adaptation can likely keep well ahead of climate change during the coming decades, with a life span of only 6 years for current crop varieties of corn, soybean, sorghum, and wheat. Advances in technology, irrespective of climate change, will lead to continuing increases in agricultural productivity in the foreseeable future (Johnson and Wittwer, 1984).

Continuing observations for the United States show that over the past 100 years, the high plains became the wheat belt during a moist period, then the dust bowl during a dry period. Agriculture through migration and technology was able to adapt. During the last century and to this writing, one sees only upward trends in production of corn, wheat, sorghum, and soybeans, attributed to new technology applied during warm and cold and wet and dry periods. American agriculture has not only persevered, but has prospered. During the past century, American farmers coped with year-to-year standard deviations of 30 ml in eastern Kansas rainfall and 2 weeks in the Minnesota growing seasons. From 1915 to 1945, Indiana farmers experienced a +0.1°C/year trend in temperatures and a total change of +2°C during the past century. The average mean temperatures from May to September in some parts of the corn and grain belt during the hot-dry summer of 1988 differed by as much as 4°C over the cool-wet year of 1992. American agriculture has already demonstrated it can adapt to a trend of +0.1°C/year, and to much greater interannual fluctuations (Benarde, 1992; Waggoner, 1983; Wittwer, 1980).

There have been continuous, both autonomous and directed, adaptations by farmers, with inputs from technology to grow crops in new environments. This is still in progress and appears to be accelerating at an increasingly rapid pace. The discovery of the New World brought the introduction of maize, potatoes, sweet potatoes, tomatoes, peanuts, cassava, cacao, as well as many kinds of peppers, beans, and squash to Europe, Asia, and Africa. These were all totally unknown outside of the Americas before the time of Columbus (McNeill, 1991). Worldwide, wheat, rice, maize, and potatoes are the four chief staples of the human diet, with sweet potatoes, peanuts, and cassava making significant contributions. Maize and potatoes had a fundamental advantage over the grains the Old World farmers already knew about. With suitable and often similar growing conditions, they produced more calories per hectare, sometimes many more. Potatoes, soybeans, wheat, rice, and maize are now successfully grown over many widely diversant climates on the earth, extending from near the arctic to the equator (Kahn, 1985).

The production of food crops, and their distribution in China, India, Africa, as well as the United States, Canada, and Europe, has undergone remarkable transitions in just the past quarter of a century. Maize for grain production has become a major crop in northwestern Europe, which was hardly existent prior to 1960. Oil crops are replacing many of the grain crops in China and long staple cotton, watermelons, corn, and peanuts are being produced in new areas with the use of plastic soil mulches. New improved varieties of rice reached a 50% adoption in only 6 years in India and the Philippines, and with wheat for Pakistan.

Thirty percent of the rice in China is now hybrid. Rapeseed (canola) has become a major new oil crop in Canada. In the United States, hybrid corn had its beginning in 1933 and achieved a 95% adoption in the state of Iowa within 7 years, and after 36 years for the entire country. Meanwhile, hybrid corn varieties have pushed the limit of corn raising in North America 500 miles to the north and winter wheat production has moved 200 miles northward (Battelle, 1973). India has now become the second largest milk producer in the world, second only to the United States, and the Chinese gooseberry has been renamed the Kiwi fruit and is being produced and marketed worldwide by New Zealand. Zimbabwe (Africa) has become a major producer of corn, and wheat production has expanded greatly in Saudia Arabia and Turkey.

A drought-tolerant sorghum variety, recently introduced in the Sudan, yields double those of traditional varieties. New cowpeas with short growing seasons, drought tolerance, and resistance to virus and bacterial diseases are available for the Sahel, and a disease-resistant cassava, with three times the yield of native strains, has been introduced into Nigeria and the adjoining lowland tropics. These are being accompanied by new polyploid cassava varieties with potential yields of 70 to 80 metric tons/ha.

The soybean, an introduction from China, was scarcely known in the United States before 1941, except as a rather insignificant soil-building and forage crop. Today, the United States produces over half the world's supply. It is one of the most significant of the world's agricultural commodities and has been one of the most lucrative of American exports. The soybean has been climatically adapted from the most northern part of Heilongjiang Province to the most southern part of Hunan in China, and from Minnesota to Louisiana in the United States. It is also grown successfully in Brazil and in the tropics of Java in Indonesia and equatorial Africa.

Most fruits and vegetables, both temperate, tropical, and subtropical, have specific climatic requirements in terms of both temperature and moisture. Mangos, papaya, lichee, bananas, and citrus, along with figs, pomegranates, and other tropical and subtropical fruits and nuts, such as the English walnut, filberts, pecans, and almond, are sensitive to low temperatures of varying lower limits. Apples, pears, and peaches require a definite number of cold units for successful fruit production. Some crops, such as the tomato, require a range of different temperatures relating to the phenological stages of flower formation, fruit setting and fruit coloration, and flavor development. Many of these temperature limitations for tree fruits and vegetables are being genetically tempered. For example, tomatoes can be successfully grown in open fields in every state, with the possible exception of Alaska in the United States, and every province in China, with some protection and at some season of the year. This is likewise true of garden beans, strawberries, radish, lettuce, cauliflower, and cabbage. Asparagus is successfully produced in the temperate climates of Michigan, New York and New Jersey, the tropics of Taiwan and China, and the Nile Delta of Egypt. Average mean daily temperatures, as well as those for night and day, vary in excess of 10°C for commercial asparagus production in the contrasting tropical growing areas of Taiwan, compared to those in the temperate zones of North America and Northern Europe. It is difficult to visualize how a global warming, even of the magnitude predicted by some model scenarios, would be catastrophic to asparagus production, and a multitude of other crops. Blueberries are successfully grown from Michigan to Florida. Apples and pears can be produced in every state in the United States, with the possible exception of Alaska, all countries in the European Community, throughout China, including Manchuria and Taiwan, in the foothills of the Himalayas in India, the highlands of Sri Lanka, and in the Nile valley of Egypt. Leaves of apricot (*Prunus armemiaca)*, a mesophilic plant, the almond and several other arid zone plants have been reported to adapt to seasonal temperature changes by shifting the temperature for maximum phytosynthetic rates upward by 10°C from spring to summer and then downward by about 10°C from summer to fall (Allen and Boote, 1992).

In summary, *the future adaptability of agricultural crop production, forests and range ecosystems, and livestock and poultry production can be indexed not only by already observed geographic distributions of culture and management but by rapid rates of change prompted, in part, by new technological inputs.* To repeat, hybrid corn production in Iowa increased from 5 to 95% of the total acreage between 1935 and 1940. Within the 1960s and 1970s, corn, arising from obscurity, became a major grain crop in northern Europe with the introduction of hybrids, and soybeans became a major crop in Brazil. The acreage of high-yielding wheat varieties in India went from nothing to 82% of the total between 1967 and 1977. Over 80% of cultivated land devoted to rice production in the Philippines is now planted to high-yielding varieties. Rice production in Indonesia doubled from 1980 to 1985. The production of sunflowers in the Red River Valley of the north in the United States, oil palm in Malaysia, and canola (rapeseed) oil production in Canada has expanded exponentially during the past two decades. Sunflowers, potatoes, sweet corn, broccoli, cabbage, onions, carrots, cauliflower, lettuce, green peas, snap beans, strawberries, blueberries, apples, and many other fruits and vegetables can be grown from Texas and Florida to Michigan,

Minnesota, and in some instances, even Alaska, and in practically every agriculturally developing country, in the Northern and Southern Hemispheres, during some season of the year. All these crops, with some technological inputs, have moved into new geographic areas within the past two decades. Some selections of dwarf apples can be successfully grown in the Nile Valley and Delta regions of Egypt, along side citrus crops. Taiwan became a major world producer of asparagus and mushrooms, normally confined to temperate climates, as tropical or semitropical crops during the 1960s, and shook world markets of these two high value food commodities (Wittwer, 1994). India experienced a literal apple revolution with their exponential expansion in production in Himachel Pradesh and Kashmir, during the 1970s.

Perhaps the most important ecosystems in the biosphere of biological productivity that will be affected by both the direct biological and climatic effects of the proposed greenhouse warming and closely allied to agriculture are the forests. As already indicated, little is known concerning either the direct biological effects of elevated above ambient CO_2 on photosynthetic and water use efficiencies or the climatic effects. The long life-span of trees, whether for food or forestry, makes their response to both climate and the direct effects of rising levels of atmospheric CO_2 very different from annual agricultural crops that constitute our basic food supply. Trees may live for hundreds of years. This suggests that they can and have withstood great climatic fluctuations and change. The forester also has the option of harvesting a tree whenever it seems appropriate. Important trees in the United States, such as the aspen, red maple, Douglas fir, and Ponderosa pine, over the latitudinal range are found from Canada to Mexico. With the highly questionable model projected changes in climate, forest succession would be modified in many parts of the world and scenarios have been drawn up as colorful illustrations. The most attention, thus far, has focused on the most northern forests of America, Europe, and Asia, rather than the mid-latitudes and the tropics. The rates of succession are fraught with uncertainties, with the time cycles probably centuries, not decades. Planted or managed forests would speed up the transition (Cooper, 1982; Reifsnyder, 1989). Schlesinger (1993a) aptly stated that to use GMCs to predict the future distribution of vegetation, as if humans were not present, is an esoteric exercise; we must remember that most natural vegetation will eventually be destroyed.

Animals play an important role in meeting the food needs in the United States, and an increasingly important role in other parts of the world, including China, with its record production of 400 million pigs, and vast increases in poultry and egg production (Wittwer et al., 1987). Likewise, India, with its water buffalos as the primary milk source, is now second only to the United States in milk production, resulting from a move of farmer's cooperatives beginning in 1970, in what has been designated as a "White Revolution" (Hoddy, 1986). The consumption of animal products has also increased precipitously in many other nations, during the 1960s and 1970s, especially in Japan, Taiwan, Korea, Israel, and the European Community (Barr, 1980). This means that livestock and poultry are now being produced in record numbers in areas not heretofore adapted for their growth and development.

Many new technological developments now emerging, and soon to be on the horizon of food-producing systems, will extend even further the geographic limits and climatic adaptability of crops and food animals in both the developed and agricultural developing world. These include the development of "transgenic" cultivars of crops of corn, cotton, soybeans, and sugar beets. One of the current introductions is a tomato, which has a vine ripe flavor with a shelf life of 14 days, and allows shipping of ready-to-eat fruit. Other soon to be introduced are cotton, corn, and potatoes that produce their own pesticides and fight off damaging insect pests, new herbicide-resistant plants that will be grown under no-till agriculture, and a variety of both high value and staple food crops that are cold, drought, and heat resistant.

A second development is the Global Positioning System (GPS). This was created by the Department of Defense and brings the "Star Wars" technology down to earth. It can be used for mapping of fields for precise and efficient fertilizer and pesticide applications in crop production, placing these inputs only where and when needed, and resulting in optimal crop productivity with minimal environmental impacts. For livestock, the most significant developments will be hormones that boost production and lower unit costs, vaccines for animal disease control, now severely limiting the geographic distribution of pigs, livestock and poultry, genetic improvements through embryo transfer, and, finally, the production of food animals and their products for special purposes.

6.5 IMPACT OF AGRICULTURAL PRACTICES ON GREENHOUSE GAS EMISSIONS AND CLIMATE CHANGE

It is reported (IPCC, 1990) that the gases that warm the surface of the earth, the so-called "greenhouse" gases, are a combination of about 1% water vapor, 0.04% CO_2, 1.72 ppm CH_4, about 300 ppb N_2O, and 280 ppb chlorofluorocarbons (CFC-11). Any direct or indirect contributions of O_3 to a greenhouse effect cannot be made with confidence because of insufficient knowledge of its distribution and behavior in the atmosphere. It does, however, have a radiative absorption potential much higher than that of CO_2. Water vapor accounts for about 60 to 70% of the warming and CO_2 for about 25%, with the balance coming from N_2O, CH_4, and CFC-11. All except water vapor have increased dramatically since 1950 (Duxbury et al., 1993). The warming induced by N_2O, CH_4, and CFC-11 is accredited to about equal that of the rising level of CO_2. Increases in atmospheric CO_2 levels have been and continue to be the main contributor to radiative forcing. They account for about 61% of the overall change and 56% in the last decade. Since 1960, the CFCs and N_2O have become progressively more important whereas the effects of CH_4 have remained about constant.

As to the anthropogenic sources of atmospheric CO_2 fossil fuel use and activities associated with its extraction and transport have contributed about 54%. Agricultural activities account for about one quarter of the total, with two thirds of that related to agricultural practices and one third associated with the conversion of land to agricultural use, mostly involving deforestation in the tropics. For estimated annual global budgets for CO_2, CH_4, and N_2O, anthropogenic activities are much more important sources of CH_4 and N_2O than they are of CO_2, accounting for 64, 24, and 3%, respectively, of total emissions (Duxbury et al., 1993). For CO_2 the chief agricultural sources arising from anthropogenic activities are vegetation and soil organic matter destruction. The chief agricultural anthropogenic sources of CH_4 arise from rice paddies, domestic ruminant animal wastes, and biomass burning; those for N_2O are from fertilizer use, grasslands, increases in cultivated lands, and biomass burning.

It is clear that there are several options to mitigate the contributions of conventional and contemporary agriculture to global greenhouse warming (Council for Agricultural Science and Technology, 1992; Duxbury et al., 1993; Kern and Johnson, 1991; Neue, 1993). Of primary importance would be a switch to conservation tillage, reduced tillage, or no-till practices. For the United States, an increase in no-till from the present 27, to 76% of the arable agricultural land would be significant since it has about 13% of the world's land under cultivation. A 10% reduction in current CH_4 emissions from rice paddies would stabilize the atmospheric load, which will arise from the projected increase in productivity and future increase in demand for this food crop. Options for mitigation include cultivar selection, nitrogen source and placement, the use of nitrification inhibitors, and alternative water management practices. Nitrous oxide mitigation strategies could focus on improved use of nitrogen fertilizer in agriculture, technologies to avoid overfertilization, greater emphasis on biological nitrogen fixation, more timely applications and placement through the use of the GPS, referred to in this chapter, and the use of nitrogen sources that minimize N_2O emissions.

Recent studies by Bronson and Mosier (1993) on N_2O emissions and CH_4 consumption in wheat and corn cropping systems emphasized that for N_2Os, which are 300 (mass basis) times more radiatively active than CO_2 and CH_4, which is about 15 times more effective, the emission of N_2O contributes far more to the "greenhouse effect" than does the consumption of CH_4 in all cropping systems. Thus, strategies to reduce N_2O emissions should be emphasized, especially in systems where CH_4 consumption rates are low.

In a further review of the effects of paddy rice culture on methane gas emissions, Lindau et al. (1993) state that rice is harvested from about 9.5% of the world's cultivated area, and that 80% of that is grown under wetland conditions. Any parameters affecting the chemical, physical, or biological characteristics of the flooded rice environment will influence methanogenesis of methane. Approximately 25% of total global methane emissions come from flooded rice paddies. Here methanogenic bacteria decompose organic molecules under anaerobic conditions. Rice plants themselves are also an important conduit for methane release (Lauren and Duxbury, 1993). Methane production in flooded rice soils is strongly influenced by soil factors, such as redox potential, temperature, pH, organic matter, soil type, and addition of nutrients and fertilizers. An understanding of these factors and related processes is important in moving to stabilize or reduce future methane emissions in flooded rice culture. The final challenge will be to develop field management practices that will both reduce methane emissions and ensure the productivity of lowland rice.

6.6 POLICY STRATEGIES RELATING TO WORLD AGRICULTURE AND THE RISING LEVEL OF GREENHOUSE GASES

There are some overall agricultural policies and measures to be endorsed relating to the conservation of resources, and maintaining high and stable food production, irrespective of a significant global warming (Bromley, 1990; Weenink, 1993). They have been frequently enumerated but not effectively implemented. Immediate action could be taken in both the short- and mid-term, starting with so called "no-regret" policies that have immediate benefits and do not conflict with probable future policy aimed at reducing greenhouse gas emissions. These short- and mid-term actions could be followed by long-term strategies based on principles, such as sustainable development and the precautionary principle to minimize the risks of global change (Weenink, 1993). Uncertainties about the rate, extent, and impact of future climate change and climatic variations should not restrain some action now. First, would be strategies toward reforestation, the conservation of energy, and a reduction of the combustion of fossil fuels, coupled with the development of renewable energy resources. For agriculture, this will include more energy-efficient farm implements and an increase in water-use efficiency, both as to energy demands and the amount of water used for crop irrigation. The current switch to conservation tillage or no-till practices, now in progress for a quarter of a century, should be accelerated. This will significantly reduce soil losses from erosion and sedimentation, conserve water resources, and increase the storage of carbon in soil organic matter, at the same time improving land productivity. It should be noted, however, that for the United States the combined savings of carbon over the next 30 years from forest plantings and changes in farming practices would be equivalent to only slightly over 1% of the projected fossil fuel use in the United States in the same period. As Duxbury et al. (1993) point out, it is clear that although some good could be achieved by changes in land management practices, alone they will do little to solve the problem of CO_2 release from fossil fuel combustion. Other practices enumerated for a main food-producing province in China are building irrigation networks, developing new cropping systems, improving varieties, programming production, and increasing family responsibility (Gao Liangzhi et al., 1989).

There are other agricultural policy strategies useful for reducing the impact of a global warming pertaining specifically to reducing emissions of nitrous oxides, accounting for 6 to 8% of the greenhouse forcing rate ascribed to anthropically derived gases; and methane, which contributes 15 to 20% of the current increase in global warming. Here, policies encouraging a change in agricultural practices would be significantly greater. Viewing policy from the standpoint of reducing sources of greenhouse emissions, there are three sources from agriculture; (1) the use of nitrogen fertilizer, (2) the cultivation of rice, and (3) enteric fermentation in domestic animals, with cattle being the major culprits. Globally, policy measures should be to reduce all three of these emissions so that the agricultural contribution to the greenhouse effect is minimal (Schuh, 1989). For the first and the third scenarios, the United States could play a key role because of the amount of nitrogen fertilizer used. For the cultivation of rice, the United States produces only 0.7% of that of the world. Nitrogen fertilizer inputs for crop production are likely to increase both for demands for more food and more land under cultivation. Conservation in the use of nitrogen and reduction in N_2O emissions may be achieved by avoiding over-fertilization, more timely applications, and continuing plant cover, so that surplus fertilizer nitrogen, that from mineralization of soil organic matter and from biological nitrogen fixation, when crops are not normally present, is recycled rather than leached, coupled with the use of slow release nitrogen sources that minimize N_2 emissions. It is believed that managed (agricultural) landscapes, in particular, temperate-region landscapes with high fertilizer inputs, and tropical landscapes undergoing large scale land conversion, are likely sources (Robertson, 1993).

A global research policy should be for improvements in fertilizer nitrogen efficiency and manure handling and procedures that reduce ammonia volatilization. The largest single energy input into world agriculture is nitrogen fertilizer. Every effort should be to improve the efficiency of its use. As an example, scarcely more than 50% of the nitrogen applied for corn production in the United States, the world's largest user of nitrogen fertilizer—half of which is used for corn production, is recovered by the crop; the remainder is lost to the environment.

As to methane emissions from ruminant animals, consideration should be given to its reduction by possibly reducing the number and size of dairy and beef cattle herds (Shuh, 1989) and further reducing methane emissions through improved genetic capacity, management, and biotechnology. The goal should be to increase efficiency and decrease methane emissions per unit of livestock product. The result should be fewer animals required to satisfy demand.

Alternatives to decreasing methane emissions from paddy rice, other than those already enumerated, could be organic residues left unincorporated (no-till), deep placement of nitrogen fertilizer, and the substitution of upland for paddy rice (Council for Agricultural Science and Technology, 1992).

A further global research policy should be to develop, genetically and through biotechnology, crops and cultivars of crops, along with cultural practices for them, and chemical treatments that would make our global food-producing system more resilient to climatic stresses and climate change. Both climate change and climatic stresses have always been present, exist now, and will prevail in the future. They are the most determinant factors in agricultural productivity and a continuing scourge to food security. The characteristics of change and variability are projections we can safely make for climate, irrespective of a global warming.

Greater resistance to climatic stresses was an important research initiative of World Food and Nutrition Study—The Potential Contributers of Research (National Academy of Sciences, 1977). It was also listed as a high priority item among the recommendations given for the next generation of agricultural research as a major national and international research priority (Wittwer, 1978a), but was never implemented.

This book has recognized and emphasized repeatedly that the most readily identifiable potential climatic impact of significant magnitude on future living standards of the human race is in agriculture. Agriculture almost everywhere is outdoors. As an industry, it is a direct recipient of the impacts of climate and weather. Climate and government policies are the major determining factors in global agricultural productivity.

Agriculture is inseparably linked to climate and the weather. In the United States, since World War II, the U.S. Department of Commerce National Weather Service has been responsible for gathering and disseminating information on weather as it relates to agriculture. The result is that the U.S. Department of Agriculture has not had a coherent weather program adapted to help farmers and ranchers cope with bad weather or exploit good weather. This has frustrated efforts to adapt agriculture to climate change, although agriculture is the nation's and the world's largest enterprise directly susceptible to changes in weather and climate. Action needs to be taken [Experiment Station Committee on Policy (ESCOP), 1994].

Because weather and climate are so critical to world agriculture and food security, there should be both a United States and world effective agricultural weather and climate network to serve farmers in minimizing losses. There should be provision for an early warning system to forecast severe storms, floods, drought, and heat and cold waves, with warnings of frost or freezing damage. An international superinformation highway is called for with easy linkages to farmers. Because of the significance of U.S. agriculture to the world economy, the widespread distribution of temperature- and moisture-recording devices and weather stations, and the climatic hazards encountered annually, it is proposed that the international leadership be in the United States.

As for policy concerning water resources, the need is here. Moreover, any degree of future global warming will likely have an impact on water and its availability for agriculture. All evidence now indicates a direct positive effect from the rising levels of atmospheric CO_2 on plant water use efficiency. As Waggoner and Revelle (1990) emphasize, we should study the effect of climate change on water resources, rather than just climate change alone. Something can be done about water resources. Agriculture is, by far, the world's greatest consumer of water. Irrigation, for example, uses about 40% of all withdrawals, and agriculture consumes fully 80 to 85% of all consumptions. Supplying water for irrigation is hard to come by, because water is used most when precipitation is light. The 17% of irrigated land of the earth provides one third of the value of crops. While climate change has not yet secured a lasting position on political policy agendas, and global warming is fraught with uncertainties, the worldwide shortages of water resources are serious, certain, already evident in many places, and have solutions. Floods and drought are the destructive extremes of the hydrologic cycle. Their negative impact on agriculture is enormous. Do we go the route of mitigation by the construction of additional water reclamation projects with benefits for agricultural production, industry, recreation, and as an energy resource with the attendant ecological disasters, destruction of natural habits, displacement of people, and loss of species? Such policy questions will not have easy answers.

Aside from the projected global water scarcity, the words of Schelling, (1983), as they appear in "Policy Implications of Greenhouse Warming" (National Academy of Sciences, 1991, 603), are appropriate: *"the most likely possibility emerging from the work done so far in relation to CO_2 is that the impact of climate change on global income and production, and especially the agriculture component of it, would not be of alarming magnitude."* Finally, from the summary of the Fifth Annual Assessment of the National Energy Outlook (National Energy Outlook Committee, 1991), we lack an understanding of the economic costs associated with various global climate change scenarios. The economic impact and costs of mitigation and adaptation are not known. A serious danger exists that we will move forward with actions to solve the "greenhouse problem" before we understand what the "problem" is all about.

7

Research Directions for the Future

7.1 THE "NO-REGRET" APPROACH

This has been emphasized repeatedly, not only in this volume, but elsewhere. It involves both policy and research. The many ingredients are outlined under Chapter 6, Section 6.6. Actions can be taken immediately, both for the short- and long-term, that have multiple benefits and will not conflict with future policy or research aimed at modifying greenhouse gas emissions and their impact on either climate change or the direct effects of elevated levels of atmospheric CO_2. Benefits would be derived whether or not there is a climate change or global warming (Weenink, 1993). These include seeking greater resilience to environmental stresses in crops and livestock and developing crop and livestock management programs for greater heat, cold, and drought resistance and protection. The development of heat–, cold-, and drought-tolerant crop varieties, with biotechnology (transgenic) input, should become an important objective for crop breeders irrespective of any global warming. Greater resilience to environmental stresses would impact favorably in food production as a means of overcoming or lessening the effects of interannual climatic variations, which have always been with us. These variations approach or exceed projected long-term climate changes. Similarly, a policy and research effort to speed up reforestation, promote soil conservation, and encourage energy conservation to increase the magnitude of the natural sinks for CO_2, and reduce consumption of fossil fuels, would be "no-regret" approaches.

There is currently no accurate listing of specific ranges of temperature of the major food crops of the earth and where production would be most favored other than in this report (Table 5.1). We need to assess the current adaptability and genetic potential for crop adaptability to ranges of climate that already exist, and those projected to exist over the surface of the earth (Rogers et al., 1992c). This assessment should first be directed to the three regions of the northern and mid-latitudes that now predominate in agricultural productivity (Chameides et al., 1994). Assessing the potential for crop adaptability to new ranges of climate now existent on the earth would be of great value in meeting world and local food needs for the future. Exploiting such potential would do much to ensure food security and maximize use of current climatic and water resources, irrespective of climate change or a global warming.

7.2 CLIMATE CHANGE

We assume that a global change in temperature and moisture levels may not be harmful and that a degree or two higher temperature, accompanied by slight precipitation increases, might be more beneficial than destructive. Also, that the present climate of the earth is not ideal and might be improved.

The experimental design would follow, in part, the assessment of Climate Change and World Food Supply in Research Report No. 3 of the Environmental Change Unit (Rosenzweig et al., 1993). From the results, we could arrive at a definition of what the optimal climate of the earth should be in terms of global food production, total biological productivity, and human health, welfare, and comfort. This is with the assumption that a warmer and wetter world would be better than what we have experienced during the past several decades.

The research would be an assessment of world agricultural productivity based on a 1 or 2°C increase in global temperature and some accompanying shifts in precipitation. There is, however, the realization that ecosystems, including agriculture, do not respond to global mean temperatures, but changes in their immediate environments. The productivity of all the major food crops would be assessed extending beyond the wheat, rice, corn, and soybeans of the Environmental Change Unit Report (Rosenzweig et al., 1993; Rosenzweig and Parry, 1994). Such a study of food crop productivity would not factor in either minor or major farm level adaptations. It would ignore any inputs of new technologies, and disregard any beneficial effects of an elevated level of atmospheric CO_2 on improved photosynthetic carbon fixation or greater water use efficiency. Who, where, and what nations and geographic areas would be the winners and losers in crop and livestock productivity?

The phenomenon of massive social, economic, and political response to perceived, dire, and unsubstantiated predictions of apocalyptic climate change in the near future is unprecedented. Scenarios include a global temperature increase ranging from 1.5 to 4.5°C within the next century. They project a climate change more rapid than any previously experienced in history. There are pronouncements of a sea level rise from thermal expansion and ice cap melting resulting in sinking of coastal cities and destruction of coastal food-producing areas. Agricultural dislocations, starvation, and civil strife resulting from ecological chaos are a common theme.

As to global warming, society is responding to a threat produced only in the world of admittedly flawed mathematical models. Any threat of climate change has been projected as a disadvantageous expectation. No real world data have been addressed to demonstrate any threat to human, animal, or plant welfare response to a CO_2 and other greenhouse gas warming.

What are the causes of the observed political, social, and economic responses, with proposals for outlays of billions of dollars to reduce CO_2 emissions, to a threat with so little supporting evidence? In the past, society has responded to threats or observations of happenings in the real world of volcanoes, earthquakes, floods, tornadoes, smelter emissions, nuclear plant malfunctions, and oil spills. Whether or not the threats materialize, what can we learn from current responses that will be of value in responding rationally to future perceived threats? Is it a matter of scientists frightening the public to get research dollars? Is it a deterioration of scientific and political credibility and integrity, coupled with beltway politics? Is it a trend of the times? Research should be directed toward this important problem (Laboratory of Climatology, Arizona State University, 1990).

7.3 THE BIOLOGICAL COMPONENT

Using a similar approach for atmospheric levels of CO_2 as for temperature (see Section 7.2), consider that the earth's overall climate is now optimal for crop and livestock production, the global economy, biological productivity and human health, welfare, and comfort. Then design models for each of the major food crops, which similators incorporate our best knowledge of how seasonal variations of temperature, light, water, nutrients and photoperiod, and interactions among them, affect essential processes of the plant and eventual productivity and yield of marketable product.

With the above in place, and with no projected climate change, add a doubling of the current ambient level of (superambient) atmospheric CO_2 and a reduction (subambient) by 30% comparable to the preindustrial level as a second variable with the ambient CO_2 level as a control comparison. The three atmospheric CO_2 levels should be maintained continuously and for the life of the crop.

The actual increase in agricultural productivity worldwide, crop by crop, can then be ascribed to the fertilizing effect of the rising level of atmospheric CO_2, now estimated to range between 5 and 10% (Allen et al., 1991a; Goudriaan and Unsworth, 1990). Projections can then be made for a doubling of current levels. Will global agricultural production increase, decrease, or remain the same? If the atmospheric accumulation of CO_2 is occurring without concomitant changes in temperature or water regimes (which to this date appear to be), then the anthropogenic increases of CO_2 could well be favorable for world agriculture, food security, total biological productivity, and the welfare of unmanaged biospheres.

Research is called for on the effects of elevated atmospheric CO_2 on the growth and physiology of plants (Rogers et al., 1992c). A decade ago, Oechel and Strain (1985) declared that less than 0.1% of the vascular plant flora had been investigated. No studies had been completed covering the life cycle of trees or other long-lived perennials. Given, however, the hundreds of thousands of plant species on the earth, the task of determining their individual CO_2 responses is overwhelming. Fortunately for agriculture and food production, we can be primarily concerned with just a few hundred, and more specifically with only 25 to 30, that stand between the survival of the human population and starvation. Perhaps the same could be done for the major tree species in the arctic, boreal, and tropical settings.

We therefore recommend using the free air CO_2 enrichment (FACE) field technology design for multiseason evaluation of crop responses, and that the studies in progress (Kimball et al., 1993b) with cotton and wheat maintained at two soil moisture and nitrogen levels be extended to the other major food and agricultural crops designated in Chapter 5 of this volume. This is to be, however, with the provision that the two levels of imposed atmospheric CO_2, the ambient and double of ambient, be maintained continuously.

A subproject would be to continue using the FACE field experiments now in progress, with cotton and wheat, to further verify the issue that there will be positive responses to the rising level of atmospheric CO_2, if other factors, such as water, light, temperatures, soil nutrients, alkalinity, air pollutants, and UV-B radiation, are limiting growth.

Similar studies, with modifications, should be extended to selected tree species and forest ecosystems and other long-lived perennials over their life cycles and to unmanaged or partially managed forest, range, and wetland communities, such as those of Drake (1992a) in the Chesapeake Bay.

Assess more thoroughly the effects of elevated levels of atmospheric CO_2 on the growth of roots and beneficial effects related thereto (Prior et al., 1994a, b; Rogers and Runion, 1994; Rogers et al., 1995). These include the enhancement of biological nitrogen fixation in legumes, certain grasses, and some forest tree species; mycorrhizal interactions for both food crops and forest trees; and other below ground processes from seed germination to maturity. This study would also relate to the following: the global productivity of root and tuber crops; the extent to which increases in biological nitrogen fixation could contribute to food production of legumes and the total output of certain forest tree species and of rangelands, where legumes are an important component; the impact on water use efficiency of plants, and improved water relations; the possibility of extending crop productivity into what are now semidesert areas; and finally, early establishment and improved survival of transplanted crop and tree seedlings. Such impacts on plant growth and improvements in food production would be of particular significance for agriculturally developing countries. If it can be established that water use efficiency of plants is significantly increased by elevated levels of atmospheric CO_2, even by a few percentage points, this could be an important, perhaps the most important effect of potential global significance, especially for the less developed world.

Conduct factorial experiments on the major food crops and tree species to separate the effects of (a) projected temperature increases; (b) the fertilizing effects of a doubling of the current atmospheric CO_2 level, (c) the projected increasing ground surface UV-B radiation, resulting from a depletion of stratospheric O_3; and (d) the damage to crops from the increasing tropospheric O_3. These should be reviewed singly and in combination with each other. According to current projections and observations for those four variables, they are simultaneously changing, relate to each other, and are co-occurring. They typify the real world in which we are now living. Emphasis should be on crop productivity and water stress (Baker and Allen, 1994).

A subproject would be to assess the possible benefits of an elevated level of atmospheric CO_2 for the amelioration of the harmful effects of air pollutants, of which tropospheric O_3 is one (Chameides et al., 1994), via stomatal closure.

A second subproject would be to assess with selective crops, reported to be most susceptible, the extent to which an elevated level (double that of current ambient) of atmospheric CO_2 would overcome the growth-retarding effects of an increase in ground level UV-B for stratospheric O_3 depletion up to 37% (approximately a doubling of ambient UV-B), and tropospheric O_3. These are levels reported to damage crops. Experimental designs could be similar to those described by Miller et al. (1994) and Krupa and Kickert (1993).

A third subproject suggested by Allen (1994) would be to assess the effect of changes in tropospheric and stratospheric O_3 on the so-called greenhouse effect impacting plant growth. UV-B radiation contributes little to the global radiation balance. Chlorofluorocarbons have recently been introduced into the earth's atmosphere and will continue for some time. They have a strong warming potential for each increment added.

A critical research effort relates to projected global warming. Effort would be directed to growth and productivity of crop plants at temperatures beyond the current high and low optima. This research is now in progress for rice and soybeans (Baker and Allen, 1993a, b). In cooler places, higher temperatures will shift biological process rates toward the optimum, and beneficial effects are likely to endure. In localities where increased temperatures move beyond the optimal, negative consequence may dominate. Will these temperature–plant growth relationships for the major food crops be likely to change in a higher CO_2 world, or specifically from a doubling of current atmospheric CO_2 levels?

Resolve the issue of down-regulation or acclimatization on photosynthetic capacity with time. Leaves of some crops (soybean, potato, cotton, citrus) maintain or even increase their photosynthetic capacity with long-term exposure to elevated atmospheric CO_2. Leaves of other crops species (rice, cabbage, kidney bean) decrease their capacity. How does this impact crop productivity (Rogers et al., 1992c)? Acclimatization is characterized by physiological changes in plants as atmospheric CO_2 increases. This has now been documented repeatedly, but usually only when rooting volumes were artificially restricted in potting of experimental plants under controlled environments or where other plant growth constraints were imposed. Such root growth and other constrictions may not characterize plants grown in well-managed biospheres, open fields, or with natural ecosystems (Johnson et al., 1993; Long and Drake, 1992). Temperature should be imposed as a variable to reaffirm Long's (1991) conclusion that there is little evidence of any feedback inhibition under warm conditions, which are generally projected to prevail with rising atmospheric levels of CO_2. Experiments should be conducted with special plant species reported to acclimate or where photosynthesis has been down-regulated with time (Long et al., 1993).

Conduct a detailed study on the effects of a doubling of current ambient levels of atmospheric CO_2 on the compositional and nutritional aspects or values of food components of the major food crops beyond that of dry matter content (Idso and Kimball, 1988). This would be significant for both people and various herbivores including livestock, poultry, wild life, and insects. Items for special attention include the following:

a. The sucrose content of sugar crops (sugarcane, sugar beet);
b. Soluble solids in grapes, citrus, and other fruits and vegetables;
c. The amount and composition of edible oils in oil crops [olive, rapeseed (canola), cotton, soybean, palm, coconut, perilla, sunflower];
d. Starch and protein content and components of root crops (potato, cassava, sweet potato, yam, cocoyam);
e. Protein and nitrogen components in cereal grains, legumes, grasses, and other forages in pastures and grazing lands;
f. Vitamins A (B-carotene) and C in numerous fruits and vegetables;
g. Effects on the milling quality of wheat and analysis of nutritional components and perishability of tropical fruits; and
h. Changes in plant/herbivore interactions with particular reference to forages for livestock and plant/insect infestations.
i. Impacts on the content of active ingredients in medicinal and pharmaceutical crops now in use for both traditional and alternative uses.

Rising levels of atmospheric CO_2 are known to favorably affect plant growth and yields of food crops. The manner in which increasing CO_2 will directly affect major plant diseases, insects, nematodes, and weed infestations, however, is fraught with many uncertainties as observed in Chapter 4 of this book. Each year tens of billions of dollars in crop yields are lost to plant pests and billions more are expended controlling them with increasing problems from chemical use involving food safety, human health, pesticide and herbicide resistance, and environmental degradation (Runion et al., 1994).

It is recommended that for the major food crops of the earth, beginning with the cereal grains, seed legumes, and root and tuber crops, the major plant pathogens, insects, and weeds for each be identified, and differences in severity of damage or infestation determined under controlled levels of infestation and at ambient and double the ambient levels of atmospheric CO_2 maintained continuously during the life of the crop.

7.4 NONMARKETABLE RESEARCH

Research should be conducted to establish values by measuring intertemporal distribution of benefits and costs of the nonmarketable effects of rising levels of atmospheric CO_2 and accompanying phenomena (National Academy of Sciences, 1992). These could include long-term health effects of the increasing ground surface ultraviolet radiation, the threat of serious degradations of global biological diversity, the loss of geopolitical position resulting from the unfavorable impact of climate change on agricultural productivity and trade, and the impact on the current and future global biological productivity as a consequence of rising levels of CO_2, tropospheric O_3, and depletion of stratospheric O_3, resulting in an increase in surface UV-B radiation.

8

Epilogue

The people of the planet earth are playing dice with their natural environment through a multitude of interventions (Nordhaus, 1994). Depending on one's perspective, it is easy to become either optimistic or pessimistic about our ability to understand and cope with the presumed threat that greenhouse warming poses to our global village. We can look at the rising level of atmospheric CO_2 as either a CO_2 problem, liability, phenomenon, subsidy, or asset.

Warnings of global warming abound (Benarde, 1992). Our natural resources of land, water, energy, and air, and natural ecosystems must be protected. People, also, must be fed from an agriculture that depends on environmental inputs, including atmospheric CO_2. This is a subsidy that is being generated in progressively greater amounts, and for which food crops are tuned to the current atmospheric influx and related environmental factors.

At the World Climate Conference (February 1979) in Geneva, Switzerland, I declared before that assembly the direct beneficial effects of the increasing level of atmospheric CO_2 on plants, and that we should not refer to it as a "CO_2 problem". Rather, it was being presumptive not knowing whether it was an asset or liability. My suggestion received a temporary element of support that did not prevail.

There has been, and still remains, a great reluctance on the part of many climatologists and ecologists, and especially environmentalists, to accept the concept that the rising level of atmospheric CO_2 could be more beneficial than harmful for plant growth, food production, and the overall biosphere (Rozema et al., 1993). Yet the scientific evidence is overwhelming. The dimunitive effects on plant growth, with emphasis on possible negative aspects to the enhanced CO_2 in a greenhouse world, has been vigorously espoused in a best seller by Gore (1992), and more recently in a special box report (World Resources Institute, 1994–95, pp. 208–209). Only five scientific reports, and a personal communication (1992) of the hundreds available, were carefully selected and repeatedly referenced. All were either negative or ignored overall beneficial results (Bazzaz and Fajer, 1992; Hileman, 1992; IPCC, 1990; Monastersky, 1993; Norby et al., 1992).

It is strange that so little research has been done during the past two decades by scientists outside the United States, Western Europe, and Australia on the direct effects of the rising level of atmospheric CO_2 on primary crop production. Worldwide attention has been directed predominantly to the negative aspects of a presumed climate change or global warming. All this seems coupled with an activism among political leaders of nations, supported by a vocal element of environmentalism, for protocols to restrict the emissions of CO_2 and other greenhouse gases into the atmosphere (Earth Summit Rio de Janiero, 1992). Such initiatives did not, initially, receive the expected support from the United States, which is labeled as the first nation and chief polluter. This has changed under the Clinton–Gore presidency. There is now an environmental initiative, in agriculture under the Presidents Climate Change Plan, to reduce greenhouse gas emissions.

Credible predictions for climate change and global food security will depend, in part, on predictions of future CO_2 concentrations in the atmosphere. Hillel and Rosenzweig (1989) state that if atmospheric CO_2 accumulation were occurring without concomitant changes in temperature and water regimes, it might be a blessing. To date, this has been true with no verifiable change in either temperature or water regimes that can be ascribed to CO_2 accumulation. CO_2 is an essential component of the vital processes of photosynthesis upon which all life on earth ultimately depends. In view of the now highly questionable model predictions of a warmer and drier climate, not vindicated thus far by any real world records, the rising atmospheric levels of CO_2 may well be a worldwide subsidy even if some warming occurs. Longer growing seasons and greater heat sums would be an asset for the productivity of most crops in temperate zone agriculture, where it is projected that the warming would most likely occur, and where most of the world's food is now produced (Chameides et al., 1994).

Many recent experiments (Allen et al., 1990) strongly suggest that CO_2 enrichments of the atmosphere engender greater plant survivability at high temperatures. Plant response to enriched CO_2 will continue to increase at high temperatures. It has been emphasized that the absolute change in growth may increase even more, or may decrease, depending on where the new higher temperature is, with respect to each plants optimal temperature. The net effect of the combined CO_2-trace gas–greenhouse effect may be well above the productivity enhancement expected from the direct fertilization effects of double the now ambient atmospheric level of 360 ppm under an unchanging climatic regime.

The rising level of atmospheric CO_2 could be the one global natural resource that is progressively increasing food production and total biological output, in a world of otherwise diminishing natural resources of land, water, energy, minerals, and fertilizer. It is a means of inadvertently increasing the productivity of farming systems and other photosynthetically active ecosystems. The effects know no boundaries and both developing and developed countries are, and will be, sharing equally. Further, there is now strong evidence that the rising abundance of this resource is enabling greater efficiency in both land and water use and enhancing the value of crop fertilization—both biological and chemical. Crop and plant responses to CO_2 enrichment offer fertile grounds, with frontiers not yet explored for enormous strides in world food production and for the conservation of water, which is rapidly becoming the world's most critical natural resource.

Dyson (1992) urged that global change research, plans, and strategies should have CO_2 effects on vegetation as a central issue rather than as a mere sideline to climate studies. Dyson (1992) further points out that the "effects of a future larger increase of atmospheric CO_2 may change the conditions of agriculture and plant ecology quite radically, long before any climate effects become apparent." A further appraisal is the comfort and convenience of the warm computer laboratory for model building—all projecting a global warming—compared to muddy, cold winter field sites, as a driving force in research choices. This level of difficulty is best recognized as a key issue in plant root research studies with CO_2. The advent of the supercomputer has accentuated the neglect of non-climatic effects (Dyson, 1992).

The uncertainties of climate and weather forecasting and, more particularly, the debates on global warming, and its magnitude, will likely continue without resolution well into the 21st century. With this in mind, the rising level of atmospheric CO_2 must be viewed with caution. The rising level is real. The principle of greenhouse gases and their growing atmospheric concentrations is certain. Changes in climate in specific fields where crops actually grow and are cultivated remain defiantly uncertain (Waggoner, 1994b). Conversely, the effects of an enriched CO_2 atmosphere on crop productivity, in large measure, are positive, leaving little doubt as to the benefits for global food security. It is, therefore, inappropriate for public discussion of the issue to focus only on the hypothetical dangers of a global warming that might result from higher CO_2 levels. It is equally important to stress the known benefits of a higher atmospheric CO_2 concentration for the production of food crops, trees, and other crops.

One of the remaining mysteries of modern science and technology, and presumably an educated and enlightened generation, is that in the majority of studies on future global food security (FAO, 1981, 1984, 1986; Meadows et al., 1972; Crosson and Anderson, 1992) there is a failure to factor in any climate variables, even though climate is the most determinant factor in agricultural productivity (Oram, 1989). Furthermore, seldom, if ever, in textbooks and other documentaries on agricultural food production, are the fertilizing effects of atmospheric CO_2 acknowledged. This was true over 30 years ago (Norman, 1962). Now, after more than a century, and with the confirmation of thousands of scientific reports, CO_2 gives the most remarkable response of all nutrients in plant bulk, is usually in short supply, and is nearly always limiting for photosynthesis. Moreover, in some of the latest reports and projections on world food production and security, the rising levels of atmospheric CO_2 as a contributing plant growth factor do not receive mention (Crosson and Anderson, 1992; Edwards et al., 1990; Per Pinstrup-Andersen, 1994; World Food Council, 1992). The rising level of atmospheric CO_2 is a universally free premium, gaining in magnitude with time, on which we all can reckon for the foreseeable future. Direct effects of increasing CO_2 on food production and the output of rangelands and forests may be more important than the effects on climate.

REFERENCES

Abelson, P. H. 1982. Improvement of Grain Crops. *Science* 216: 4543.
Abelson, P. H. 1989a. Climate and Water. *Science* 243: 461.
Abelson, P. H. 1989b. World Food Research. *Science* 244: 125.
Abelson, P. H. 1990. Uncertainties About Global Warming. *Science* 247: 1529.
Abelson, P. H. 1992. Agriculture and Climate Change. *Science* 257: 9.
Acock, B. 1990. Effects of Carbon Dioxide on Photosynthesis, Plant Growth, and Other Processes. In: *Impact of Carbon Dioxide, Trace Gases, and Climate Change on Global Agriculture.* B. A. Kimball et al. (Eds.), pp, 45–60. American Society of Agronomy Special Publication Number 53, Madison, WI.
Acock. B., and L. H. Allen, Jr. 1985. Crop Response to Elevated Carbon Dioxide Concentrations. In: *Direct Effects of Increasing Carbon Dioxide on Vegetation.* B. R. Strain and J. D. Cure (Eds.), pp. 53–97, U.S. Department of Energy, DOE/ER-0238, Washington, D.C.
Acock, B., and D. Pasternak. 1986. Effects of CO_2 Concentration on Composition, Anatomy, and Morphology of Plants. In: *Carbon Dioxide Enrichment of Greenhouse Crops,* Vol. II. H. Z. Enoch and B. A. Kimball (Eds.), pp. 41–52. CRC Press, Boca Raton, FL.
Adams, M. B., and G. E. Taylor, Jr. 1990. Effects of Ozone on Forests in the Northeastern United States, In: *Ozone Risk Communication and Management.* E. J. Calabrese, C. E. Gillert, and B. D. Beck, (Eds.), pp. 65–92. Lewis Publishers, Chelsea, MI.
Adams, R. M., S. A. Hamilton, and B. A. McCarl. 1984. The Economic Effects of Ozone on Agriculture. Environmental Research Laboratory. Environmental Protection Agency, Corvallis, OR.
Adams, R. M., J. D. Glyer, and B. A. McCarl. 1989. The Economic Effects of Climate Change on U.S. Agriculture: A Preliminary Assessment. In: *The Potential Effects of Global Climate Change on the United States.* Appendix C. Agriculture Volume 1, pp. 4.1–4.56. Environmental Protection Agency, Washington, D.C.
Adams, R. M., C. Rosenzweig, R. M. Peart, J. T. Ritchie, B. A. McCarl, J. D. Glyer, R. B. Cury, J. W. Jones, K. J. Boote, and L. H. Allen, Jr. 1990. Global Climate Change and U.S. Agriculture. *Nature (London)* 345: 219–224.
Akey, D. H., and B. A. Kimball. 1989. Growth and Development of the Beet Armyworm on Cotton Grown in Enriched Carbon Dioxide Atmosphere. *Southwest Entom.* 14: 255–260.
Akey, D. H., B. A. Kimball, and J. R. Mauney. 1988. Growth and Development of the Pink Bollworm, *Pectinophora gossypiella* (Lepidoptera: Gelechiidae) on Bolls of Cotton Grown in Enriched Carbon Dioxide Atmospheres. *Environ. Entomol.* 17: 452–455.
Allen, L. H., Jr. 1979. Potentials for Carbon Dioxide Enrichment. In: *Modification of the Aerial Environment of Crops.* B. J. Barfield and J. F. Gerber (Eds.), pp. 500–519. Am. Soc. Agr. Eng., St. Joseph, MI.
Allen, L. H., Jr. 1990. Plant Response to Rising Carbon Dioxide and Potential Interactions with Air Pollutants. *J. Environ. Qual.* 19(1): 15–34.
Allen, L. H., Jr. 1991. Effects of Increasing Carbon Dioxide Levels and Climate Change on Plant Growth, Evapotranspiration, and Water Resources, In: *Managing Water Resources in the West Under Conditions of Climate Uncertainty,* pp. 101–147. Proceedings of a Colloquium, November 14–16, 1990. Scottsdale, AZ. National Academy Press, Washington, D.C.

Allen, L. H., Jr. 1994. UV Radiation as Related to the Greenhouse Effect. In: NATO ASI series, Vol. 118. *Stratospheric Ozone Depletion/UV–B Radiation in the Biosphere.* R. H. Biggs and M. E. B. Joyner (Eds.). Springer–Verlag, Berlin.

Allen, L. H., Jr., P. Jones, and J. W. Jones. 1986. Rising Atmospheric CO_2 and Evapotranspiration. American Society of Agricultural Engineers. Publication 14-85, pp. 13–27, St. Joseph, MI.

Allen, L. H., Jr., K. J. Boote, J. W. Jones, P. H. Jones, R. R. Valle, B. Acock, H. H. Rogers, and R. C. Dahlman. 1987. Response of Vegetation to Rising Carbon Dioxide: Photosynthesis, Biomass, and Seed Yield of Soybean. In: *Global Biogeochemical Cycle.s* 1: 1–14.

Allen, L. H. Jr., K. J. Boote, P. H. Jones, J. W. Jones, A. J. Rowland–Bamford, G. Bowes, D. A. Graetz, and K. R. Reddy. 1989a. Temperature and CO_2 Effects on Rice: 1988. U.S. Department of Energy, Office of Energy Research, Carbon Dioxide Research Division, Washington, D.C.

Allen, L. H., Jr, R. M. Peart, J. W. Jones, R. B. Curry, and K. J. Boote. 1989b. Likely Effects of Climate Change Scenarios on Agriculture of the USA. In: *Preparing for Climate Change.* J. C. Topping (Ed.), pp. 186–191. The Climate Institute, Washington, D.C.

Allen, L. H., Jr., E. C. Bisbal, K. J. Boote, and P. H. Jones. 1991a. Soybean Dry Matter Allocation under Subambient and Superambient Levels of Carbon Dioxide. *Agron. J.* 83(5): 875–883.

Allen, L. H., Jr., K. J. Boote, J. W. Jones, P. H. Jones, J. T. Baker, S. L. Albrecht, R. S. Wasehmann, and F. Kamuru. 1991b. *Response of Vegetation to Carbon Dioxide. Carbon Dioxide Effects on Growth, Photosynthesis and Evapotranspiration of Rice at Three Nitrogen Fertilizer Levels.* U.S. Department of Energy, Carbon Dioxide Research Division and U. S. Department of Agriculture, Number 062.

Allen, L. H., Jr., and K. J. Boote. 1992. Vegetation, Effects of Rising CO_2. In: *Encyclopedia of Earth Systems,* Vol. 4, pp. 409–416. Academic Press, New York.

Allen, R. G., and F. N. Gichuki. 1989. Effects of Projected CO_2-Induced Climatic Changes on Irrigation Water Requirements in the Great Plains States (Texas, Oklahoma, Kansas, and Nebraska). In: *The Potential Effects of Global Climatic Change on the United States.* Appendix C. *Agriculture,* Vol. 1, pp. 6.1–6.42. Environmental Protection Agency, Washington, D.C.

Allen, S. G., S. B. Idso, B. A. Kimball, and M. G. Anderson. 1988. Relationship Between Growth Rate and Net Photosynthesis of Azolla in Ambient and Elevated CO_2 Concentrations. *Agric. Ecosyst. Environ.* 20: 137–141.

Allen, S. G., S. B. Idso, B. A. Kimball, J. T. Baker, L. H. Allen, Jr., J. R. Mauney, J. W. Radin, and M. G. Anderson. 1990. *Effects of Air Temperature on Atmospheric CO_2–Plant Growth Relationships.* U.S. Department of Energy, Carbon Dioxide Research Program and U.S. Department of Agriculture, DOE/ER-0450T, Washington, D.C.

American Society of Agronomy. 1990. *Impact of Carbon Dioxide, Trace Gases, and Climate Change on Global Agriculture.* ASA Special Publication Number 53. Madison, WI.

Ames, D. 1980. Thermal Environmental Affects Production Efficiency of Livestock. *BioScience* 30(7): 457–460.

Amthor, J. S. 1991. Respiration in a Future, Higher–CO_2 World. *Plant, Cell Environ.* 14: 13–20.

Anonymous. 1993a. Biodiversity: There's a Reason for It. *Science* 262: 1511.

Anonymous, 1993b. Look to the Clouds. *Dev. Cooperation Mag.,* February: 56–62.

Arp, W. J. 1991 Effects of Source–Sink Relations on Photosynthetic Acclimation to Elevated CO_2. *Plant, Cell Environ.* 14: 869–875.

Arp, W. J., and B. G. Drake. 1991. Increased Photosynthetic Capacity of *Scirpus olneyi* after 4 Years of Exposure to Elevated CO_2. *Plant, Cell Environ.* 14: 1003–1006.

Arp, W. J., B. G. Drake, W. T. Pockman, P. S. Curtis, and D. F. Whigham. 1993. Interactions Between C_3 and C_4 Salt Marsh Plant Species During Four Years Exposure to Elevated Atmospheric CO_2. *Vegetatio* 104/105: 133–143.

Arrhenius, S. 1896. On the Influence of Carbonic Acid in the Air upon Temperature of the Ground . *Phil. Mag.* 41: 237–275.

Arteca, R. N., B. W. Poovaiah, and O. E. Smith. 1979. Changes in Carbon Dioxide Fixation, Tuberization and Growth Induced by CO_2 Application to the Root Zone of Potato Plants. *Science* 205: 279–280.

Atmospheric Environment Service of Canada. 1994. *Modelling the Global Climate System.* Climate and Atmospheric Research Directorate, Downsview, Ontario.

Austin, R. B. 1989. The Climate Vulnerability of Wheat. In: *Climate and Food Security,* pp, 123– 135. International Rice Research Institute , Manila, Philippines and American Society for the Advancement of Sciences, Washington, D.C. 1989.

Ausubel, J. H. 1983. Historical Note. *Changing Climate.* National Research Council, Washington, D. C.

Ausubel, J. H. 1991a. A Second Look at the Impacts of Climate Change. *Am. Sci.* 70: 210–221.

Ausubel, J. H. 1991b. Does Climate Still Matter? *Nature* 350: 649–652.

Avery, D. T. 1991. *Global Food Progress.* The Hudson Institute, Indianapolis, IN.

Bacastow, R. B., C. D. Keeling, and T. C. Whort. 1985. Seasonal Amplitude Increase in atmospheric CO_2 Concentration at Mauna Loa, Hawaii. 1959–1982. *J. Gcophys. Res.* 90 (D6); 10,529–10,540.

Bach, W. 1979. The Impact of Increasing Atmospheric CO_2 Concentrations on the Global Climate: Potential Consequences and Corrective Measures. *Environ. Int.* 3: 215–228.

Bach, W. 1994. A Climate and Environmental Protection Strategy. The Road Toward a Sustainable Future. *Climate Change* 27(2): 147–160.

Baker, J. T., and L. H. Allen, Jr. 1993a. Effects of CO_2 and Temperature on Rice: A Summary of Five Growing Seasons. *J. Agr. Met.* 48(5): 575–582.

Baker, J. T., and L. H. Allen, Jr. 1993b. Contrasting Crop Species Responses to CO_2 and Temperature: Rice, Soybean and Citrus. *Vegetatio* 104/105: 239–260.

Baker, J. T., and L. H. Allen, Jr. 1994. Assessment of the Impact of Rising Carbon Dioxide and Other Potential Climate Changes on Vegetation. *Environ. Poll.* 83: 223–235.

Baker, B. B., J. D. Hanson, R. M. Bourdon, and J. B. Eckert. 1993. The Potential Effects of Climate Change on Ecosystem Processes and Cattle Production. In: *U.S. Rangelands. Climate Change* 25: 97–117.

Baker, D. N., and H. Z. Enoch. 1983. Plant Growth and Development. In: *CO_2 and Plants. The Response of Plants to Rising Levels of Atmospheric Carbon Dioxide,* A Selected Symposium No. 84. E. R. Lemon (Ed.), pp. 107–130. Westview Press, Boulder, CO .

Baker, J. T., L. H. Allen, Jr., and K. J. Boote. 1990. Growth and Yield Responses of Rice to Carbon Dioxide Concentration. *J. Agr. Sci.* 115: 313–320.

Baker, J. T., L. H. Allen, Jr., and K. J. Boote. 1992a. Temperature Effects on Rice at Elevated CO_2 Concentration. *J. Exp. Bot.* 43:(252): 959–964.

Baker, J. T., L. H. Allen, Jr., and K. J. Boote. 1992b. Response of Rice to Carbon Dioxide and Temperature. *Agr. Forest Meteorol.* 60: 153–160.

Baker, J. T., F. Laugel, K. J. Boote, and L. H. Allen, Jr. 1992c. Effects of Daytime Carbon Dioxide Concentration on Dark Respiration of Rice. *Plant, Cell Environ.* 15: 231–239.

Balling, R. C., Jr. 1992. *The Heated Debate. Greenhouse Predictions versus Climate Reality.* Pacific Research Institute for Public Policy, San Francisco, CA.

Balling, R.C., Jr. 1993. The Global Temperature Data. In: *Research and Exploration.* 9(2): 201–207. National Geographic Society, Washington, D.C.

Barr, T. N. 1980. The World Food Situation and Global Grain Prospects. *Science* 214: 1087–1095.

Bauerle, W. L., D. W. Kretchman, and L. L. Tucker-Kelly. 1986. CO_2 Enrichment in the U.S. In: *Carbon Dioxide Enrichment of Greenhouse Crops,* Vol. I. H. Z. Enoch and B. A. Kimball, (Eds.), pp, 49–57. CRC Press, Boca Raton, FL.

Battelle. 1973. *Interactions of Science and Technology in the Innovative Process: Some Case Studies.* Final Report Prepared for National Science Foundation, Columbus, OH.

Bazzaz, F. A. 1990. The Response of Natural Ecosystems to Rising Global CO_2 Levels. *Annu. Rev. Ecology and System.* 21: 167–196.

Bazzaz, F. A., K. Garbutt, E. G. Reekie, and W. E. Williams. 1989. Using Growth Analysis to Interpret Competition between a C_3 and a C_4 Annual under Ambient and Elevated CO_2. *Oecologia* 79: 223–235.

Bazzaz, F. A., and E. D. Fajer. 1992. Plant Life in a CO_2-Rich World. *Sci. Am.* 266(1): 68–74.

Beer, S. 1986. The Fixation of Inorganic Carbon in Plant Cells. In: *Carbon Dioxide Enrichment of Greenhouse Crops,* Vol. II. H. Z. Enoch, and B. A. Kimball (Eds.), pp. 3–11. CRC Press, Boca Raton, FL.

Benarde, M. A. 1992. *Global Warning–Global Warming.* John Wiley, New York.

Besford, R. T. 1993. Photosynthetic Acclimation in Tomato Plants Grown in High CO_2 *Vegetatio* 104/105: 441–448.

Bhattacharya, N. C., P. K. Biswas, S. Bhattacharya, N. Sionit, and B. P. Strain. 1985a. Growth and Yield Response of Sweet Potato (Ipomoea batatas) to Atmospheric CO_2 Enrichment. *Crop Sci.* 25: 975–981.

Bhattacharya, S., N. C. Bhattacharya, and B. R. Strain. 1985b. Rooting of Sweet Potato Stem Cuttings under CO_2-Enriched Environment and with IAA Treatment. *Hort. Sci.* 20: 1109–1110.

Bhattacharya, N. C., D. R. Hileman, P. P. Ghosh, R. L. Musser, S. Bhattacharya, and P. K. Biswas. 1990. Interaction of Enriched CO_2 and Water Stress on the Physiology of and Biomass Production in Sweet Potato Grown in Open-Top Chambers. *Plant, Cell Environ.* 13: 933–940.

Billings, W. D. 1970. *Plants, Man, and the Ecosystem.* Wadsworth, Belmont, CA.

Billings, W. D., K. M. Peterson, J. O. Luken, and D. A. Mortensen. 1984. Interaction of Increasing Atmospheric Carbon Dioxide and Soil Nitrogen on the Carbon Balance of Tundra Microcosms. *Oecologia* 65: 26–29.

Black, C. C., Jr. 1986. Effects of CO_2 Concentration on Photosynthesis and Respiration of C4 and CAM plants. In: *Carbon Dioxide Enrichment of Greenhouse Crops,* Vol. II. H. Z. Enoch and B. A. Kimball (Eds.), pp. 29–40. CRC Press, Boca Raton, FL.

Black, V. J. 1982. Effects of Sulphur Dioxide on Physiological Processes in Plants. In: *Effects of Gaseous Air Pollution in Agriculture and Horticulture,* M. H. Unsworth and D. P. Ormrod (Eds.), pp. 67–91. Buttersworth, London.

Blackman, F. F. 1905. Optima and Limiting Factors. *Am. Bot.* 19: 281–295.

Blasing, T. J., and A. M. Solomon. 1982. Response of the North American Corn Belt to Climatic Warming. Publ. 2134. Environ. Sci. Div., Oak Ridge Natl. Lab., Oak Ridge, TN.

Boyer, J. S. 1982. Plant Productivity and Environment. *Science* 218: 443–448.

Bradley, R. S. 1988. The Explosive Volcanic Eruption Signal in Northern Hemisphere Continental Temperature Records. *Climate Change* 12: 221–243.

Bravdo, B. A. 1986. Effect of CO_2 Enrichment on Photosynthesis of C_3 Plants. In: *Carbon Dioxide Enrichment of Greenhouse Crops.* H. Z. Enoch, and B. A. Kimball (Eds.), pp. 13–27. CRC Press, Boca Raton, FL.

Bretherton, F. P., K. Bryan, and J. D. Woods. 1990. Time-dependent Greenhouse Gas Induced Climate Change. In: *Climate Change: The IPCC Scientific Assessment.* J. T. Houghton et al. (Eds.), pp. 179–193. Cambridge University Press, Cambridge.

Bromley, D. A. 1990. The Making of Greenhouse Policy. *Issues Sci. Technol.,* 7(1): 55–69.

Bronson, K. F., and A. R. Mosier. 1993. Nitrous Oxide Emissions and Methane Consumption in Wheat and Corn-Cropped Systems in Northeastern Colorado. In: *Agricultural Ecosystem Effects on Trace Gases and Global Climate-Change,* pp. 133–144. American Society of Agronomy Publication No. 55, Madison, WI.

Brooks, W. T. 1989. The Global Warming Panic. *Forbes* December, 96–102.

Brun, W. A., and R. L. Cooper. 1967. Response of Soybeans to a Carbon Dioxide-Enriched Atmosphere. *Crop Sci.* 7: 455–467.

Bryson, R. A. 1993. Simulating Past and Forecasting Future Climates. *Environ. Conservation* 20: 339–345.

Budyko, M. I. 1982. *The Earth's Climate: Past and Future.* Academic Press, New York.

Bunce, J. A. 1992. Light, Temperature and Nutrients as Factors in Photosysnthetic Adjustment to an Elevated Concentration of Carbon Dioxide. *Physiol. Plant.* 86: 173–179.

Butler, G. D. 1985. Populations of Several Insects on Cotton in Open-Top Carbon Dioxide Enrichment Chambers. *Southwest. Entomol.* 10:264–267.

Butler, G. D., B. A. Kimball, and J. R. Mauney. 1986. Populations of Bemisia tobaci (Homoptea: Aleyrodidae) on Cotton Grown in Open-Top Field Chambers Enriched with CO_2 *Environ. Entomol.* 15: 61–63.

Campbell, W. J., L. H. Allen, Jr., and G. Bowes. 1990. Response of Soybean Canopy Photosynthesis to CO_2 Concentration, Light and Temperature. *J. Exp. Bot.* 41: 427–433.

Carlson, R. W., and F. A. Bazzaz. 1980. The Effects of Elevated Carbon Dioxide Concentrations on Growth, Photosynthesis, Transpiration, and Water Use Efficiency of Plants. In: *Environmental and Climatic Impact of Coal Utilization,* J. J. Singh, and A. Dupack (Eds.), pp. 609–622, Academic Press, New York.

Carlson, R. W., and F. A. Bazzaz. 1982. Photosynthetic and Growth Responses to Fumigation With SO_2 for C_3 and C_4 Plants. *Oecologia* 54: 50–54.

Cathey, H. M., and H. E. Heggestad. 1982. Ozone and Sulfur Dioxide Sensitivity of Petunia; Ozone Sensitivity of Herbaceous Plants; Ozone Sensitivity of Woody Plants: Modification by Ethylenediurea. *J. Am. Hort. Sci.* 107(6): 1028–1045.

Cess, R. D., M. H. Zhang, G. L. Potter et al., 1993. Uncertainties in Carbon Dioxide Radiative Forcing in Atmospheric General Circulation Models. *Science* 262: 1252–1255.

Chameides, W. L., P. S. Kasibhatta, J. Yienger, and H. Leny II. 1994. Growth of Continental-Scale Metro-Agro-Plexes, Regional Ozone Pollution, and World Food Production. *Science* 264: 74–77.

Changnon, S. A. 1992. Inadvertent Weather Modification in Urban Areas: Lessons for Global Climate Change. *Bull. Am. Meteorol. Soc.* 73(5): 619–627.

Chapman, H. W., and W. E. Loomis. 1953. Photosynthesis in the Potato under Field Conditions. *Plant Physiol.* 28: 703–716.

Chapman, H. W., L. S. Gleason, and W. E. Loomis. 1954. The Carbon Dioxide Content of Field Air. *Plant Physiol.* 29: 500–503.

Chaudhuri, U. N., R. B. Burnett, M. B. Kirkham, and E. T. Kanemasu. 1986. Effect of Carbon Dioxide on Sorghum Yield, Root Growth and Water Use. *Agr. For. Meteorol.* 37: 109–122.

Chaudhuri, U. N., M. B. Kirkham, and E. T. Kanemasu. 1990. Root Growth of Winter Wheat under Elevated Carbon Dioxide and Drought. *Crop Sci.* 30: 853–857.

China National Rice Research Institute (CNRRI). 1991. Prospects of Rice Farming for 2000. *Proceedings of the CNRRI Inauguration Ceremony and Rice Research Conference,* Oct. 9–11, 1989, Hangzhou, China. Zhejiang Publishing House of Science and Technology, Hangzhou, China.

Clawson, M. 1979. Forests in the Long Sweep of American History. *Science* 204: 1168–1174.

Cock, J. H. 1982. Cassava: A Basic Energy Source in the Tropics. *Science* 218: 755–762.

Cock, J. H., and S. Yoshida. 1973. Photosynthesis, Crop Growth and Respiration of Tall and Short Rice Varieties. *Soil Sci. Plant Matr.* 19: 53–59.

Collins, W. B. 1976. Effect of Carbon Dioxide Enrichment on Growth of the Potato Plant. *HortScience* 11: 467–469.

Cooper, C. F. 1982. Food and Fiber in a World of Increasing Carbon Dioxide. In: *Carbon Dioxide Review 1982.* W. C. Clark (Ed.), pp. 299–320. Oxford University Press, New York.

Cooper, R. L., and W. A. Brun. 1967. Response of Soybeans to Carbon Dioxide-Enriched Atmosphere. *Crop Sci.* 7: 455–457.

Council for Agricultural Science and Technology. 1992. *Preparing U. S. Agriculture for Global Climate Change.* Council of Agricultural Science and Technology, Ames, IA.

Cowling, E. B. 1987. *Improving the Stewardship of Our Natural Resources and Environment in Agricultural Research for a Better Tomorrow. Commemorating the Hatch Act Centennial 1887–1997,* p. 111. U. S. Department of Agriculture, Washington, D.C.

Crosson, P., and J. R. Anderson. 1992. *Resources and Global Food Prospects.* World Bank Technical Paper 184. The World Bank, Washington, D.C.

Cui, M., and P. S. Nobel. 1994. Ges Exchange and Growth Response to Elevated CO_2 and Light Levels in the CA M species, *Opuntia fiscus-indica. Plant, Cell Environ.* 17: 935–944.

Cure, J. D. 1985. Carbon Dioxide Doubling Responses: A Crop Survey. In: *Direct Effects of Increasing Carbon Dioxide on Vegetation,* B. R. Strain and J. D. Cure (Eds.), pp. 99–110. U.S. Department of Energy, DOE/ER–0238, Washington, D.C.

Cure, J. D., and B. Acock. 1986. Crop Response to Carbon Dioxide Doubling: A Literature Survey. *Agr. For. Meteorol.* 38: 127–145.

Curtis, P. S., B. G. Drake, and D. F. Whigham. 1989. Nitrogen and Carbon Dynamics in C_3 and C_4 Estuarine Marsh Plants Grown under Elevated CO_2 Sites. *Oecologie* 78: 297–301.

Dahlman, R. C. 1993. CO_2 and Plants: Revisited. *Vegetatio* 104/105: 339–355.

Dahlman, R. C., J. F. Reynolds, and B. R. Strain. 1986. Modeling the Response of Ecosystems to Fossil Fuel Emissions: Carbon Dioxide and Pollutant Interactions. *Air Poll. Control Assoc. Proc.* 86-9.5: 1–15.

Darrall, N. W. 1989. The Effect of Air Pollutants on Physiological Processes in Plants. *Plant, Cell Environ.* 12: 1–30.

Decker, W. L., V. Jones, and R. Ochutuni. 1985. The Impact of CO_2-Induced Climate Change on U.S. Agriculture. In: *Characterization of Information Requirements for Studies of CO_2 Effects: Water Resources, Agriculture, Fisheries, Forests and Human Health,* pp. 69–93. U.S. Department of Energy, DOE/ER-0236, Washington, D.C.

Del Castillo, D., B. Acock, V. R. Reddy, and M. C. Acock. 1989. Elongation and Branching of Roots on Soybean Plants in a Carbon Dioxide-Enriched Aerial Environment. *Agron. J.* 81: 692–695.

D'Itri, F. (Ed.). 1982. *Acid Precipitation, Effects on Biological Systems.* Ann Arbor Science. The Butterworth Group, Ann Arbor, MI.

Dobson, A. 1992. Withering Heats. *Nat. Hist.* 101(9): 2–8.

DOE Multi-Laboratory Climate Change Committee Report. 1990. *Energy and Climate Change.* Lewis Publishers, Chelsea, MI.

Downing, T. E. 1992. *Climate Change and Vulnerable Places: Global Food Security and Country Studies in Zimbabawe, Kenya, Senegal and Chile.* Environmental Change Unit, University of Oxford, Oxford, England.

Downton, W. J. S., O. Bjorkman, and C. S. Pike. 1980. Consequences of Increased Atmospheric Concentrations of Carbon Dioxide for Growth and Photosynthesis of Higher Plants. In: *Carbon Dioxide and Climate,* G. I. Pearman (Ed.), pp. 143–151. Australian Academy of Science, Canberra.

Drake, B. G. 1992a. The Impact of Rising CO_2 on Ecosystem Production. In: *Natural Sinks of* CO_2. J. Wisniewski, and A. E. Lugo (Eds.), pp. 25–44. Kluwer Academic Publishers, Boston.

Drake, B. G. 1992b. A Field Study of the Effects of Elevated CO_2 on Ecosystem Processes in a Chesapeake Bay Wetland. *Aust. J. Bot.* 40: 579–595.

Drake, B. G. 1992c. The Impact of Rising CO_2 on Ecosystem Production. *Water, Air Soil Poll.* 64: 25–44.

Drake, B. G., and R. C. Dahlman. 1994. The Potential Effect of Rising Atmospheric CO_2 Concentration on Carbon Assimilation in Terrestrial Ecosystems. 4th International CO_2 Conference, Sept. 15, Carqueiranne, France.

Drake, B. G., I. H. Ziska, J. A. Bunce, W. J. Arp, K. Hogan, and A. P. Smith. 1995. Dark Respiration in Plants Grown in Field Exposed to Elevated Atmospheric CO_2. *Plant, Cell Environ.* 18: (in press).

Dudek, D. J. 1989. Climate Change Impacts upon Agriculture and Resources: A Case Study of California. In: *The Potential Effects of Global Climate Change on the United States.* Appendix C. Agriculture Volume I, pp. 5.1–5.38. Environmental Protection Agency, Washington, D.C.

Durning, A. T. 1993. *Saving the Forests. What Will It Take?* Worldwatch Paper 117. Worldwatch Institute, Washington, D. C.

Duxbury, J. M., L. A. Harper, and A. R. Mosier. 1993. Contributions of Agroecosystems to Global Climate Change. In: *Agricultural Ecosystem Effects on Trace Gases and Global Climate Change,* pp. 1–18. American Society of Agronomy Publication Number 55, Madison, WI.

Dyson, F. 1992. Carbon Dioxide in the Atmosphere and Biosphere. In: *From Eros to Gaia.* pp. 130–148. Pantheon Books, New York.

Eamus, D., and P. G. Jarvis. 1989. The Direct Effects of Increase in the Global Atmospheric CO_2 Concentration on Natural and Commercial Temperate Trees and Forests. *Adv. Ecol. Res.* 19: 1–55.

Eastin, J. D. 1980. Sorghum. In: *Potential Productivity of Field Crops under Different Environments,* pp. 187–204. International Rice Research Institute, Manila.

Edwards, C. A., R. Lal, P. Madden, R. H. Miller, and G. House. 1990. *Sustainable Agricultural Systems.* Soil and Water Conservation Society, Ankeny, IA.

Egli, D. B., J. W. Pendleton, and D. B. Peters. 1970. Photosynthetic Rate of Three Soybean Communities as Related to Carbon Dioxide Levels and Solar Radiation. *Agron. J.* 62: 411–414.

Electric Power Research Institute. 1990. The Science of Global Warming. *EPRI J.* 15(4): 4–13.

Ellsaesser, H. W. 1993. Global Warming, A Different View. Paper Presented at the ECO Conference, 18–20 February, 1993. Reno Hilton Hotel, Reno, NV.

El–Sharkaway, M. A. 1993. Drought Tolerant Cassava for Africa, Asia and Latin America. *BioScience* 43(7): 441–451.

Elston, J. 1980. Climate. In: *Potential Productivity of Field Crops under Different Environments,* pp. 1–14. International Rice Research Institute, Manila.

Emmert, E. M. 1955. *Low-Cost Plastic Greenhouses.* Kentucky Agr. Exp. Sta. Progress Report 29, Lexington.

Enoch, H. Z., and B. A. Kimball (Eds.). 1986. *Carbon Dioxide Enrichment of Greenhouse Crops. II. Physiology, Yield, and Economics.* Vols. 1 and 2. CRC Press, Boca Raton, FL.

Erwin, J. E., R. D. Heins, and M. G. Karlsson. 1989. Thermomorphogenesis in *Lilium longiflorum. Am. J. Bot.* 76(1): 47–52.

Evans, L., R. Petterson, H. S. J. Lee, and P. G. Jarvis. 1993. Effects of Elevated CO_2 on Birch. In: *CO_2 and Biosphere.* J. Rozema et al. (Eds), pp. 452–453. Kluwer Academic Publishers, Boston.

Essex, C. 1986. Trace Gases and the Problem of False Invariants in Climate Models—a Comment. *Climatol. Bull.* 20(1): 19–25.

Experiment Station Committee on Policy (ESCOP). 1994. Agricultural Experiment Stations and Global Climate Change. Cooperative State Research Service and the University of Georgia Exp. Sta., Griffin.

Fajer, E. D., M. D. Bowers, and F. A. Bazzaz. 1989. The Effects of Enriched Carbon Dioxide Atmospheres on Plant–Insect Herbivore Interactions. *Science* 243: 1198–1200.

FAO, Food and Agricultural Organization of the United Nations. 1981. *Agriculture Toward 2000.* FAO, Rome.

FAO, Food and Agricultural Organization of the United Nations. 1984. *Land, Food and People.* FAO, Rome.

FAO, Food and Agricultural Organization of the United Nations. 1986. Food Crop and Shortages, *Global Information and Early Warning System on Food and Agriculture.* Special Report, FAO, Rome.

Felker, P., and B. S. Bandurski. 1979. Uses and Potential Uses of Leguminous Trees for Minimal Energy Input Agriculture. *Econ. Bot.* 33: 172–174.

Fischer, R. A., and M. I. Aguilar. 1976. Yield Potential in a Dwarf Spring Wheat and the Effect of Carbon Dioxide Fertilization. *Agron. J.* 68: 749–752.

Fiscus, E. L., J. E. Miller, and F. L. Booker. 1995. Is UV-B a Hazard to Soybean Photosynthesis and Yield? Results of an Ozone/UV-B Interaction Study and Model Predictions. In: *Stratospheric Ozone Depletion/UV-B Radiation in the Biosphere,* R. H. Biggs (Ed.), pp. 135–147. Springer-Verlag, Berlin (in press).

Fisher, K. S., and A. F. E. Palmer. 1983. Maize. In: *Potential Productivity of Field Crops under Different Environments,* pp. 155–180. International Rice Research Institute, Manila.

Flavin, C. 1989. *Slowing Global Warming: A Worldwide Strategy.* Worldwatch Paper 91. Washington, D.C.

Follett, R. F. 1993. Global Climate Change, U. S. Agriculture, and Carbon Dioxide. *J. Prod. Agr.* 6: 181–190.

Ford, M. A., and G. N. Thorne. 1967. Effect of CO_2 Concentration on Growth of Sugarbeet, Barley, Kale, and Maize. *Ann. Bot.* 31: 629–694.

Friedly, J. 1993. Taking the Heat. *Palo Alto Weekly* March 24: 26–29.

Gaastra, P. 1959. Photosynthesis of Crop Plants as Influenced by Light, Carbon Dioxide, Temperature and Stomatal Diffusion Resistance. *Meded. Landbouwhogesch.* Wageningen 59: 1–68.

Gao Liangzhi, Juan Fang, and Li Bingbai. 1989. Climatic Variation and Food Production in Jiangsu, China. In: *Climate and Food Security,* pp. 557–562. International Rice Research Institute, Manila, and American Association for the Advancement of Science, Washington, D.C.

Garcia, R. L., S. B. Idso, and B. A. Kimball. 1993. *CO_2 Enrichment of Pine Trees.* Annual Report. U.S. Water Conservation Laboratory, USDA, Phoenix, AZ.

Garcia, R. L., S. B. Idso, G. W. Wall, and B. A. Kimball. 1994. Changes in net Photosynthesis and Growth of *Pinus elderica* Seedlings in Response to Atmospheric CO_2 Enrichment. *Plant, Cell Environ.* 17: 971–978.

Gates, D. M. 1985. Global Biospheric Response to Increasing Atmospheric Carbon Dioxide Concentration. In: *Direct Effects of Increasing Carbon Dioxide and Vegetation,* B. R. Strain and J. D. Cure (Eds.), pp. 171–184. U.S. Department of Energy, DOE/ER-0238. Washington, D.C.

Gifford, R. M. 1974. A Comparison of Potential Photosynthesis, Productivity and Yield of Plant Species with Differing Photosynthetic Metabolism. *Aust. J. Plant Physiol.* 1: 107.

Gifford, R. M. 1977. Growth Pattern, Carbon Dioxide Exchange, and Dry Weight Distribution in Wheat Growing under Differing Photosynthetic Environments. *Aust. J. Plant Physiol.* 4: 99–110.

Gifford, R. M. 1979a. Growth and Yield of CO_2-Enriched Wheat under Water-Limited Conditions. *Aust. J. Plant Physiol.* 6: 367–378.

Gifford, R. M. 1979b. Carbon Dioxide and Plant Growth under Water and Light Stress: Implications for Balancing the Global Carbon Budget. *Search* 10: 316–318.

Gifford, R. M. 1988. Directs Effects of Higher Carbon Dioxide Concentrations on Vegetation. In: *Greenhouse: Planning for Climate Change,* G. I. Pearman, (Ed.), pp. 506–519. E. J. Brill, New York.

Gifford, R. M. 1989. Exploiting the Fertilizer Effect of Increasing Atmospheric Carbon Dioxide. In: *Climate and Food Security,* pp. 477–487. International Rice Research Institute, Manila and American Society for Horticultural Science, Washington, D.C.

Gifford, R. M., H. Lambers, and J. I. L. Morison. 1985. Respiration of Crop Species under CO_2 Enrichment. *Plant Physiol.* 63: 351–356.

Goldsberry, K. L. 1986. CO_2 Fertilization of Carnations and Some Other Flower Crops. In: *Carbon Dioxide Enrichment of Greenhouse Crops,* Vol. 2, H. Z. Enoch, and B. A. Kimball (Eds.), pp. 117–140. CRC Press, Boca Raton, FL.

Gore, A. 1992. *Earth in the Balance.* Houghton Mifflin, New York.

Goudriaan, J., and R. J. Bijlsma. 1987. Effect of CO_2-Enrichment on Growth of Faba Beans at Two Levels of Water Supply. *Netherlands J. Agr. Sci.* 35: 189–191.

Goudriaan, J., and M. H. Unsworth. 1990. Implications of Increasing Carbon Dioxide and Climate Change for Agricultural Productivity and Water Resources. In: *Impact of Carbon Dioxide, Trace Gases, and Climate Change on Global Agriculture,* pp. 111–130. ASA Special Publication. No. 53. American Society of Agronomy, Madison, WI.

Goudriaan, J., H. van Keulen, and H. H. van Laar (Eds.). 1990. *The Greenhouse Effect and Primary Productivity in European Agro-Ecosystems.* Pudoc, Wageningen.

Graham, R. C., M. C. Turner, and V. H. Dale. 1990. How Increasing CO_2 and Climate Change Affect Forests. *BioScience* 40(8): 575–587.

Graumlich, L. J. 1991. Subalpine Tree Growth, Climate, and Increasing CO_2: An Assessment of Recent Growth Trends. *Ecology* 71: 1–11.

Green, K., and R. D. Wright. 1977. Field Responses of Photosynthesis to CO_2 Enhancement in Ponderosa Pine. *Ecology* 58: 687–692.

Graybill, D. A. and S. B. Idso. 1993. Detecting the Aerial Fertilization Effect of Atmospheric CO_2 Enrichment in Tree-Ring Chronologies. *Global Biogeochem. Cycles* 7: 81–95.

Grulke, N. E., G. H. Reichers, W. C. Oechel, U. Hjelm, and C. Jaeger. 1990. Carbon Balance in Tussock Tundra under Ambient and Elevated Atmospheric CO_2 *Oecologia* 83: 485–494.

Guilderson, T. P., R. G. Fairbanks, and J. L. Rubenstone. 1994. Topical Temperature Variations Since 20,000 Years Ago. Modulating Interhemispheric Climate Change. *Science* 263: 663–665.

Gunderson, C. A., R. J. Norby, and S. D. Wullschleger. 1993. Foliar Gas Exchange Responses of Two Deciduous Hardwoods During 3 Years of Growth in Elevated CO_2: No Loss of Photosynthetic Enhancement. *Plant, Cell Environ.* 16: 797–807.

Hahn, G. L. 1976. Rational Environmental Planning for Efficient Livestock Production. *Biometeorology* 6: 106–114.

Hahn, S. K. 1977. Sweet Potato. In: *Ecophysiology of Tropical Crops,* R. T. Alvim and T. T. Kozlowaski (Eds.), pp. 237–248. Academic Press, New York.

Hahn, S. K. 1993. *The Genetic Basis of Cassava Improvement in Nigeria. Past Work and Future Prospects.* International Institute of Tropical Agriculture, Ibadan, Nigeria.

Hahn, S. K., and Y. Hozyo. 1983. Sweet Potato and Yam. In: *Potential Productivity of Field Crops under Different Environments,* pp. 319–340. International Rice Research Institute, Manila.

Hanan, J. J. 1986. CO_2 Enrichment for Greenhouse Rose Production. In: *Carbon Dioxide Enrichment of Greenhouse Crops,* Vol. 2, H. Z. Enoch, and B. A. Kimball (Eds.), pp. 141–149. CRC Press, Boca Raton, FL.

Hansen, J., A. Lacis, and M. Prather. 1989. Greenhouse Effect of Chlorofluorocarbons and Other Trace Gases. *J. Geophys. Res.* 94: 16417–16421.

Hansen, J., A. Lacis, R. Ruedy, M. Sato, and H. Wilson. 1993. How Sensitive Is the World's Climate? In: *Research and Exploration,* Vol. 9(2), pp. 142–158. National Geographic Society, Washington, D.C.

Hanson, H., N. E. Borlaug, and R. G. Anderson. 1982. *Wheat in the Third World.* Westview Press, Boulder, CO.

Hardman, L. L., and W. A. Brun. 1971. Effects of Atmospheric Carbon Dioxide Enrichment at Different Developmental Stages on Growth and Yield Components of Soybeans. *Crop Sci.* 11: 886–888.

Hardy, R. W. F., and U. D. Havelka. 1975. Nitrogen Fixation Research: A Key to World Food? *Science* 188: 633–643.

Hardy, R. W. F., and U. D. Havelka. 1977. Possible Routes to Increase the Conversion of Solar Energy to Food and Feed by Grain Legumes and Cereal Grains (Crop Production): CO_2 and N_2 Fixation. In: *Biological Solar Energy Conversion.* A. Mitsui, S. Miyachi, A. San Pietro, and S. Tamura (Eds.), pp. 299–322, Academic Press, New York.

Hare, F. K. 1980. Climate and Agriculture. The Uncertain Future. *J. Soil Water Conserv.* 35(3): 112–115.

Hare, F. K. 1983. The Build-up of CO_2 (Book Review of W. C . Clark, Ed.). Carbon Dioxide Review. 1982. Oxford University Press. *Science* 219–283.

Hartt, C. F., and G. O. Burr. 1967. Factors Affecting Photosynthesis in Sugarcane. *Proc. Int. Soc. Sugar Cane Technol.* 12: 590–609.

Havelka, U. D., V. A. Wittenback, and M. G. Boyle. 1984. CO_2-Enrichment Effects on Wheat Yield and Physiology. *Crop Sci.* 24: 1163–1168.

Heck, W. W. 1989. Assessment of Crop Losses from Air Pollutants in the United States. In: *Air Pollution Toll on Forests and Crops.* J. J. Mackenzie, and M. T. El-Ashry (Eds.), pp. 235–315. Yale University Press, New Haven, CT.

Heck, W. W., O. C. Taylor, R. Adams, G. Bingham, J. Miller, E. Preston, and L. Weinstein. 1982. Assessment of Crop Loss from Ozone. *J. Air Poll. Control Assoc.* 32(4): 353–361.

Heck, W. W., O. C. Taylor, and D. T. Tingey (Eds.). 1988a. *Assessment of Crop Loss from Air Pollutants.* Elsevier Applied Science, New York.

Heggestad, H. E ., T. J. Gish, E. H. Lee, J. H. Bennett, and L. W. Douglas. 1985. Interactions of Soil Moisture Stress and Ambient Ozone on Growth and Yield of Soybeans. *Phytopathology* 75: 472–477.

Heinz–Ulrich Neue. 1993. Methane Emission from Rice Fields. *BioScience* 47(7): 466–473.

Hendrey, G. R., K. F. Levin, and J. Nagy. 1993. Free Air Carbon Dioxide Enrichment: Development, Progress, Results. *Vegetatio* 104/105: 17–31.

Hicklenton, P. R. 1988. *CO_2 Enrichment in the Greenhouse.* Growers Handbook Series Vol 2, Timber Press, Portland, OR.

Hillel, D., and C. Rosenzweig. 1989. *The Greenhoouse Effect and its Implications Regarding Global Agriculture.* Massachusetts Agr. Exp. Sta. Res. Bull. 724. Amherst, MA.

Hileman, B. 1992. Web of Interactions Makes It Difficult to Untangle Global Warming Data. *Chem. Eng. News* 70(117): 13.

Hoddy, E. 1986. India's White Revolution. *D + C Mag.* 4/86: 19–22.

Hoffman, J. S. 1984. Carbon Dioxide and Future Forests. *J. For.* 82: 164–167.

Hoffman, J. S., S. Seidel, J. O'Callahan, R. Sandenburgh, J. L. Kirk, E. Taylor, M. Gibbs, E. Marshall, M. Modlin, C. Markert, J. Tableporter, and E. Pechan. 1982. *Strategic Options for Enhancing Forest Industry Productivity in the Face of Rising Atmospheric Carbon Dioxide.* ICF, Inc. and the U.S. Environmental Protection Agency, Washington, D.C.

Holm–Hansen, O. 1981. *Effect of Increased CO_2 on Ocean Biota.* Special Report, University of California, La Jolla.

Hoogendoorn, J. 1985. The Physiology of Variation in the Time of Ear Emergence among Wheat Varieties from Different Regions of the World. *Euphytica* 34: 559–571.

Hopen, H. J., and S. K Ries. 1962. The Mutually Compensating Effect of Carbon Dioxide Concentrations and Light Intensities on the Growth of *Cucumis sativus L. Proc. Am. Soc. Hort. Sci.* 81: 358–364.

Hou, L., A. C. Hill, and A. Soleimani. 1977. Influence of CO_2 on the Effects of SO_2 and NO_2 on Alfalfa. *Environ. Poll.* 12: 7–16.

Houghton, R. A., and G. M. Woodwell. 1989. Global Climate Change. *Sci. Am.* 260: 36–44.

Hunt, R., D. W. Hand, M. A. Hannah, and A. M. Neal. 1991. Response to CO_2 Enrichment in 27 Herbaceous Species. *Funct. Ecol.* 5: 410–421.

Idso, K. E. 1993. Scientific Paper No. 23, Arizona State University, Laboratory of Climatology, Tempe, AZ.

Idso, K. E., and S. B. Idso. 1994. Plant Response to Atmospheric CO_2 Enrichment in the Face of Environmental Constraints: A Review of the past 10 years. *Agr. For. Meteorol.* 69: 153–203.

Idso, S. B. 1980. The Climatological Significance of a Doubling of the Earth's Atmospheric Carbon Dioxide Concentration. *Science* 207: 1462–1463.

Idso, S. B. 1984. Carbon Dioxide and Climate: Is There a Greenhouse in Our Future? *Quart. Rev. Biol.* 59: 291–294.

Idso, S. B. 1985. The Search for Global Greenhouse Effects. *Environ. Conserv.* 12: 29–35.

Idso, S. B. 1989a. *Carbon Dioxide and Global Change: Earth in Transition.* IBR Press, Tempe, AZ.

Idso, S. B. 1989b. Three stages of Plant Response to Atmospheric CO_2 Enrichment. *Plant Pysiol. Biochem.* 27(1): 131–134.

Idso, S. B. 1990. Interactive Effects of Carbon Dioxide and Climate Variables in Plant Growth. In: *Impact of Carbon Dioxide, Trace Gases, and Climate Change on Global Agriculture,* pp. 61–69. ASA Publication No. 53. American Society of Agronomy, Madison, WI.

Idso, S. B., B. A. Kimball, and M. G. Anderson. 1986. Temperature Increases in Water Hyacinth Caused by Atmospheric CO_2 Enrichment. *Arch. Meteorol. Geophys. Biochim. Ser. B* 36: 365–370.

Idso, S. B., B. A. Kimball, and J. R. Mauney. 1987. Atmospheric Carbon Dioxide Enrichment Effects on Cotton Midday Foliage Temperature: Implications for Plant Water Use and Crop Yield. *Agron. J.* 79: 667–672.

Idso, S. B., and B. A. Kimball. 1988. Atmospheric CO_2 Enrichment and Plant Dry Matter Content. *Agr. For. Meteorol.* 43: 171–181.

Idso, S. B., S. G. Allen, and B. A. Kimball. 1990. Growth Response of Water Lily to Atmospheric CO_2 Enrichment. *Aquat. Bot.* 37: 87–92.

Idso, S. B., and B. A. Kimball. 1991. Downward Regulation of Photosynthesis and Growth at High CO_2 Levels. *Plant Physiol.* 96: 990–992.

Idso, S. B., B. A. Kimball, and S. G. Allen. 1991. Net Photosynthesis of Sour Orange Trees Maintained in Atmospheres of Ambient and Elevated CO_2 Concentrations. *Agr. For. Meteorol.* 54: 95–101.

Idso, S. B., and B. A. Kimball. 1992. Effects of Atmospheric CO_2 Enrichment on Photosynthesis, Respiration, and Growth of Sour Orange Trees. *Plant Physiol.* 99: 341–343.

Idso, S. B., and B. A. Kimball. 1993. Tree Growth in Carbon Dioxide Enriched Air and Its Implications for Global Carbon Cycling and Maximum Levels of Atmospheric CO_2. *Global Biogeochem. Cycles* 7: 537–555.

Imai, K., and D. F. Coleman. 1983. Elevated Atmospheric Partial Pressure of Carbon Dioxide and Dry Matter Production of Konjak (*Amorphophallus konjai* K. Koeh). *Photosynthesis Res.* 4: 331–336.

Imai, K., D. F. Coleman, and T. Yanagisawa. 1984. Elevated Atmospheric Partial Pressure of Carbon Dioxide and Dry Matter Production of Cassava (*Manihot esculenta,* Crantz). *Jpn. J. Crop Sci.* 53: 479–485. See also *Jpn. J. Crop Sci.* 54: 413–418.

Intergovernmental Panel on Climate Change (IPCC). 1990. *The IPCC Scientific Assessment Report.* Prepared by Working Group I. J. T. Houghton, G. J. Jenkins, and J. J. Ephraums (Eds.). Cambridge University Press, Cambridge, U.K.

Intergovernmental Panel on Climate Change (IPCC). 1992. *Climate Change 1992. The Supplementary Report to the IPCC. Scientific Assessment.* J. T. Houghton, C J. Jenkins, and J. J. Ephraums (Eds.). Cambridge University Press, Cambridge, U.K.

Intergovernmental Panel on Climate Change (IPCC). 1993. *Scientific Assessment of Climate Change.* WMO/UNDP. Environmental Change Unit. University of Oxford, Oxford, U.K.

International Rice Research Institute (IRRI) and American Association for the Advancement of Science (AAAS). 1989. *Climate and Food Security.* International Symposium on Climate Variability and Food Security in Developing Countries, 5–9 February 1987, New Delhi, India, Washington, D.C., and Manila.

International Rice Research Institute (IRRI). 1993. *Program Report for 1992,* pp. 52–53. Manila.

International Rice Research Institute (IRRI). 1994. Personal Correspondence. Manila.

Irvine, J. E. 1990. Sugarcane. In: *Potential Productivity of Field Crops Under Different Environments,* pp. 361–381. International Rice Research Institute, Manila.

Jarvis, P. G., J. L. Monteith, W. J. Shuttleworth, and M. H. Unsworth. 1989. Forests, Weather and Climate. *Phil. Transact. Roy. Soc. London, (Series B).* 324: 369–392.

Jin Zhiging, Zheng Xilian, Fang Juan, Ge Dackuo, and Chen Hue. 1992. *Potential Effects of Global Climate Change on Winter Wheat Production in China.* Jiangsu Academy of Agricultural Sciences, Nanjing, China.

Johnson, G. L., and S. H. Wittwer. 1984. *Agriculture Technology until 2030: Prospects, Priorities, and Policies.* Michigan State University Agr. Exp. Sta. Special Report 12. East Lansing.

Johnson, H. B., H. W. Polley, and H. S. Mayeux. 1993. Increasing CO_2 and Plant–Plant Interactions: Effects on Natural Vegetation. *Vegetatio* 104/105: 157–170.

Jones, P., L. H. Allen, Jr., and J. W. Jones. 1985a. Responses of Soybean Canopy Photosynthesis and Transpiration to Whole-Day Temperature Changes in Different CO_2 Environments. *Agron. J.* 77: 242–249.

Jones, P., L. H. Allen, Jr., J. W. Jones, and R. Valle. 1985b. Photosynthesis and Transpiration Responses of Soybean Canopies to Short-Term and Long-Term CO_2 Treatments. *Agron. J.* 77: 119–126.

Jones, P. D., and T. M. L. Wigley. 1990. Global Warming Trends. *Sci. Am.* 259: 84–91.

Jones, P. D. et al. 1990. Assessment of Urbanization Effects in Time Series of Surface Air Temperatures over Land. *Nature (London)* 347: 169–172.

Kahn, E . J., Jr. 1985. *The Staffs of Life.* Little Brown, Boston.

Kane, S., J. Reilly, and J. Tobey. 1992. An Imperial Study of the Economic Effects of Climate Change on World Agriculture. *Climate Change* 21: 17–35.

Karl, T. R. 1993. Missing Pieces of the Puzzle. In: *Research and Exploration,* Vol. 9(2), pp. 234–249. National Geographic Society, Washington, D.C.

Karl, T. R., G. Kukla, V. N. Razuvayev, M. C. Changery, R. G. Quayle, R. R. Heim, Jr., D. R. Esterling, and C. B. Fu. 1991. Global Warming: Evidence for Asymmetric Diurnal Temperature Change. *Geophys. Res. Lett.* 18: 2253–2256.

Karl, T. R., P. D. Jones, R. W. Knight, G. Kukla, N. Plummer, V. Razuvayev, K. P. Gallo, J. Lidseay, R. J. Charlson, and T. C. Peterson. 1993. A New Perspective on Recent Global Warming: Asymmetric Trends of Daily Maximum and Minimum Temperature. *Bull. Am. Meteord. Soc.* 74(6): 1007–1023.

Katz, R. W., and B. G. Brown. 1992. Extreme Events in a Changing Climate: Variability Is More Important Than Averages. *Climate Change* 21: 289–302.

Kauppi, P. E., K. Mielikainen, and K. Kuusela. 1992. Biomass and Carbon Budget of European Forests, 1971 to 1990. *Science* 256: 70–74.

Keeling, C. D. 1983. *The Global Carbon Cycle: What We Know and Could Know from Atmospheric Biospheric and Oceanic Observations,* pp. II. 3–II.62. CONF–820970, U.S. Department of Energy, Washington, D.C.

Keeling, C. D., R. B. Bacastow, R. B. Carter, S. C. Piper, T. P. Whorf, M. Heimann, W. G. Mook, and H. Roeloffzen. 1989. A Three Dimensional Model of Atmospheric CO_2 Transport Based on Observed Winds: Observational Data and Preliminary Analysis. In: *Aspects of Climate Variability in the Pacific and Western Americas.* D. H. Peterson (Ed.), pp. 165–236. Geophysical Monograph 55. American Geophysical Union, Washington, D.C.

Kellogg, W. W. 1991a. Overview of Global Environmental Change: The Science and Social Science Issue. *MTS J.* 25(3): 5–11.

Kellogg, W. W. 1991b. Response to Skeptics of Global Warming. *Bull. Am. Meteorol. Soc.* 72(4): 499–511.

Kellogg, W. W., and R. Schware. 1981. *Climate Change and Society: Consequences of Increasing Atmospheric Carbon Dioxide.* Westview Press, Boulder, CO.

Kenny, G. J., P. A. Harrison, and M. L. Parry (Eds.). 1993. *The Effect of Climate Change on Agricultural and Horticultural Potential in Europe.* Environmental Change Unit, University of Oxford, Oxford.

Kern, J. S., and M. G. Johnson. 1991. *The Impact of Conservation Tillage Use on Soil and Atmospheric Carbon in the Contiguous United States.* EPA/600/3-91-056. USEPA Environ. Res. Lab., Corvallis, OR.

Kerr, R. A. 1993a. Pinatubo Global Cooling on Target. *Science* 259: 594.

Kerr, R. A. 1993b. The Ozone Hole Reaches a New Low. *Science* 262: 501.

Kerr, R. A. 1994a. Methane Increase Put on Pause. *Science* 263: 751.

Kerr, R. A. 1994b. Did Pinatubo Send Climate Warming Gases Into a Dither? *Science* 263: 1562.

Kerr, J. B., and C. T. McElroy. 1993. Evidence for Large Upward Trends of Ultraviolet–B Radiation Linked to Ozone Depletion. *Science* 262: 1032–1034.

Khanna-Chopra, R., and S. K. Sinha. 1989. Impact of Climate Variation on Production of Pulses. In: *Climate and Food Security,* pp. 219–236. International Rice Research Institute, Manila and American Association for the Advancement of Science, Washington, D.C.

Kickert, R. N., and S. V. Krupa. 1990. Forest Responses to Tropospheric Ozone and Global Climate Change. An Analysis. *Environ. Poll.* 68: 29–65.

Kimball, B. A. 1983a. Carbon Dioxide and Agricultural Yield: An Assemblage and Analysis of 430 Prior Observations. *Agron. J.* 75: 779–788.

Kimball, B. A. 1983b. *Carbon Dioxide and Agricultural Yield. An Assemblage and Analysis of 770 Prior Observations.* WCL Report 14. U.S. Water Conservation Laboratory, USDA, Phoenix, AZ.

Kimball, B. A. 1985. Adaptation of Vegetation and Management Practices to a Higher Carbon Dioxide World. In: *Direct Effects of Increasing Carbon Dioxide on Vegetation,* B. R. Strain, and J. D. Cure (Eds.), pp. 185–204, U. S. Department of Energy, DOE/ER-0238, Washington, D.C.

Kimball, B. A. 1986a. CO_2 Stimulation of Growth and Yield under Environmental Constraints. In: *Carbon Dioxide Enrichment of Greenhouse Crops,* Vol. 2, H. Z. Enoch and B. A. Kimball (Eds.), pp. 53–67 CRC Press, Boca Raton, FL.

Kimball, B. A. 1986b. Influence of Elevated CO_2 on Crop Yield. In: *Carbon Dioxide Enrichment of Greenhouse Crops,* Vol. 2, H. Z. Enoch and B. A. Kimball (Eds.), pp. 105–115. CRC Press, Boca Raton, FL.

Kimball, B. A., and S. B. Idso. 1983. Increasing Atmospheric CO_2: Effects on Crop Yield, Water Use, and Climate. *Agr. Water Manage.* 7: 55–72.

Kimball, B. A., G. H. Heichel, C. W. Stuber, D. E. Kissel, and S. Ernst (Eds.). 1990a. *Impact of Carbon Dioxide, Trace Gases, and Climate Change on Global Agriculture.* ASA Special Publication Number 53. American Society of Agronomy, Madison, WI.

Kimball B. A., N. J. Rosenberg, and L. H. Allen, Jr. Eds. 1990b. *Impact of Carbon Dioxide, Trace Gases and Climate Change on Global Agriculture.* ASA Special Publication 53. American Society of Agronomy, Madison, WI.

Kimball, B. A., J. R. Mauney, F. S. Nakayama, and S. B. Idso. 1993a. Effects of Elevated CO_2 and Climate Variables in Plants. *J. Soil Water Conserv.* 48: 9–14. See also Effects of Increasing Atmospheric CO_2 on Vegetation. *Vegetatio* 104/105: 65–75.

Kimball, B. A., et al. 1993b. *Carbon Dioxide Enrichment: Data on the Response of Cotton to Varying CO_2, Irrigation and Nitrogen.* Oak Ridge National Laboratory, Environmental Sciences Division, Publication No. 3880 (ORNL/CDIAC-44 NDP-037). Oak Ridge, TN.

King, K. M., and D. H. Greer. 1986. Effects of Carbon Dioxide Enrichment and Soil Water on Maize. *Agron. J.* 78:515–521.

Knapp, A. K., J. T. Fahnestock, and C. E. Owensby. 1994. Elevated Atmospheric CO_2 Alters Stomatal Responses to Variable Sunlight in a C_4 Grass. *Plant, Cell Environ.* 17:189–195.

Korner, C., and J. A. Arnone III. 1992. Responses to Elevated Carbon Dioxide in Artificial Tropical Ecosystems. *Science* 257: 1672–1675.

Kramer, P. J. 1981. Carbon Dioxide Concentration, Photosynthesis and Dry Matter Production. *BioScience* 31: 29–33.

Krenzer, E. G., Jr., and D. N. Moss. 1975. Carbon Dioxide Enrichment Effect upon Yield and Yield Components of Wheat. *Crop Sci.* 15: 71–74.

Krizek, D. T. 1975. Influence of Ultraviolet Radiation on Germination and Early Seedling Growth. *Physiol. Plant.* 34: 182–186.

Krupa, S. V., and W. J. Manning. 1988. Atmospheric Ozone: Formation and Effects on Vegetation. *Environ. Poll.* 50: 101–137.

Krupa, S. V., and R. N. Kickert. 1989. The Greenhouse Effect: Impacts of Ultraviolet-B (UV-B) Radiation, Carbon Dioxide (CO_2) and Ozone (O_3) on Vegetation. *Environ. Poll.* 61: 263–392.

Krupa, S. V., and R. N. Kickert. 1993. The Greenhouse Effect: The Impacts of Carbon Dioxide (CO_2), Ultraviolet-B (UV-B) Radiation and Ozone (O_3) on Vegetation (Crops). *Vegetatio* 104/105: 223–238.

Krupa, S. V., and M. Nosal. 1989. Effects of Ozone on Agricultural Crops. In: *Atmospheric Ozone Research and Its Policy Implications,* T. Schneider, S. D. Lee, G. J. Wolters, and L. D. Grant (Eds.), pp. 229–238. Elsevier, Amsterdam.

Krupa, S. V., H. H. Rogers, and G. B. Runion. 1993. Varying Crop Response to Increasing CO_2. A Written Statement Submitted to The Enquete-Kommission Protecting the Earth's Atmosphere, German Bundestag, Bonn 1, Germany.

Kukla, G., T. R. Karl, and M. C. Riches (Eds.). 1994. Asymmetric Change of Temperature Range. U.S. Department of Energy 025, Washington, D.C.

Laboratory of Climatology, Arizona State University. 1990. Proceedings of a Research Symposium, Oct. 12–13. Research Agenda. Global Climatic Change: A New Vision for the 1990's. Vols. 1, 2. Tempe, AZ.

Laing, D. R., P. J. Kretchmer, S. Zuluaga, and P. G. Jones. 1983. Field Bean. In: *Potential Productivity of Field Crops under Different Environments,* pp. 227–248. International Rice Research Institute, Manila.

Lambers, H. 1993. Rising CO_2, Secondary Plant Metabolism, Plant–Herbivore Interactions and Litter Decomposition. *Vegetatio* 104/105: 263–271.

Lamborg, M. R., R. W. F. Hardy, and E. A. Paul. 1983. Microbial Effects. In: *CO_2 and Plants. The Response of Plants to Rising Levels of Atmospheric Carbon Dioxide,* E. R. Lemon (Ed.), pp. 131–176. AAAS Selected Symposium No. 84. Westview Press, Boulder, CO.

Landsberg, H. E. 1984. Global Climate Trends. In: *The Resourceful Earth, A Response to Global 2000.* Simon, J. L. and H. Kahn, (Eds.), pp. 272–315. Basil Blackwell, New York.

Lauren, J. G., and J. M. Duxbury. 1993. Methane Emissions from Flooded Rice Amended with a Green Manure. In: *Agricultural Ecosystem Effects on Trace Gases and Global Climate Change,* pp. 183–192. ASA Publication No. 55. American Society of Agronomy, Madison, WI.

Lawlor, D. W., and R. A. C. Mitchell. 1991. The Effects of Increasing CO_2 on Crop Photosynthesis and Productivity: A Review of Field Studies: A Commissioned Review. *Plant, Cell Environ.* 14: 807–818.

Layser, E. F. 1980. Forestry and Climate Change. *J. For.* 78: 678–682.

Leadley, P. W., and B. G. Drake. 1993. Open Top Chambers for Exposing Plant Canopies to Elevated CO_2 Concentration and for Measuring Net Gas Exchange. *Vegetatio* 104/105: 3–15.

Lemon, E. R. 1977. The Lands Response to More Carbon Dioxide. In: *The Fate of Fossil Fuel CO_2 in the Ocean.* N. R. Anderson and A. Malahoff (Eds.), pp. 97–130. U.S. Office of Naval Research, Symposium Series in Oceanography, Plenum, New York.

Lemon, E. R. (Ed.). 1983. *CO_2 and Plants. The Response of Plants to Rising Levels of Atmospheric Carbon Dioxide.* AAAS Selected Symposium 84. Westview Press, Boulder, CO.

Lincoln, D. E. 1993. The influence of Plant Carbon Dioxide and Nutrient Supply on Susceptibility of Insect Herbivores. *Vegetatio* 104/105: 273–280.

Lincoln, D. E., N. Sionit, and B. R. Strain. 1984. Growth and Feeding Responses of *Pseudoplusia includens* (Lepidoptera-Noctuidae) to Host Plants Grown in Controlled Carbon Dioxide Atmospheres. *Environ. Entomol.* 13: 1527–1530.

Lincoln, D. E., D. Couvet, and N. Sionit. 1986. Responses of an Insect Herbivore to Host Plants Grown in Carbon Dioxide Enriched Atmospheres. *Oecologia* 69: 556–560.

Lindau, C. W., W. H. Patrick, Jr., and R. D. De Laune. 1993. Factors Affecting Methane Production in Flooded Rice Soils. In: *Agricultural Ecosystem Effects on Trace Gases and Global Climate Change,* pp. 157–165. ASA Publication No. 55. American Society of Agronomy, Madison, WI.

Lindzen, R. S. 1990. Some Coolness Concerning Global Warming. *Bull. Am. Meteorol. Soc.* 71: 288–299.

Lindzen, R. S. 1991. Response to AMS Policy Statement on Global Warming Change. *Bull. Am. Meteorol. Soc.* 72: 515.

Lindzen, R. 1993. Absence of Scientific Basis. In: *Research and Exploration,* Vol. 9(2), pp. 191–200. National Geographic Society, Washington, D.C.

Liscio, J. 1993. Pinatubo and the Prairie. A Pacific Volcano Slams the Corn Belt. *Barrons* LXXIII(34): 8–9, 20–21.

Long, S. P. 1991. Modification of the Response of Photosynthetic Productivity to Rising Temperature by Atmospheric CO_2 Concentrations: Has Its Importance Been Underestimated? *Plant, Cell Environ.* 14: 729–739.

Long, S. P., and B. G. Drake. 1992. Photosynthetic CO_2 Assimilation and Rising Atmospheric CO_2 Concentration. In: *Crop Photosynthesis: Spatial and Temporal Determinants,* N. R. Baker, and H. Thomas (Eds.), pp. 69–103. Elsevier Science Publishers, Amsterdam.

Long, S. P., N. R. Baker, and C. A. Raines. 1993. Analyzing the Response of Photosynthetic CO_2 Assimilation to Long–Term Elevation of Atmospheric CO_2 Concentration. *Vegetatio* 104/105: 33–45.

Lorenzo, P., C. Maroto, and N. Castilla. 1990. CO_2 in Plastic Greenhouse in Almeria. *Acta Horticult.* 268: 165–169.

Luxmoore, R. J., E. G. O'Neill, J. M. Ells, and H. H. Rogers. 1986. Nutrient Uptake and Growth Responses of Virginia Pine to Elevated Atmospheric Carbon Dioxide. *J. Environ. Qual.* 15(3): 244–251.

MacCracken, M. C., M. I. Budyko, A. D. Hecht, and Y. A. Izrael. 1990. *Prospects for Future Climate.* A Special US/USSR Report on Climate and Climate Change. Lewis Publishers, Chelsea, MI.

MacDonald, G. J. F., R. Revelle, G. M. Woodwell, and C. D. Keeling. 1979. A Warning on Synfuels, CO_2 and the Weather. *Science* 205: 376–377.

MacKenzie, D. 1985. Ethiopia: Famine Amid Genetic Plenty. New Sci., August 8: 22–23.

Majernik, O., and T. A. Mansfield. 1972. Stomatal Responses to Atmospheric CO_2 Concentrations During Exposure of Plants to SO_2 Pollution. *Environ. Poll.* 3: 1–7.

Mandels, M., J. Nystrom, and L. A. Spano. 1973. *Enzymatic Hydrolysis of Cellulosic Wastes. Fact Sheet.* U. S. Army Natick Laboratories, Natick, MA.

Mangelesdorf, P. C. 1986. The Origin of Corn. *Sci. Am.* 255: 80–86.

Manning, W. J. 1990. Effects of Ozone on Crops in New England. In: *Ozone Risk Communication and Management.* E. J. Calabrese, C. E. Gilbert, and B. D. Beck (Eds.), pp. 57–63. Lewis Publishers, Chelsea, MI.

Martin, P., N. J. Rosenberg, and M. S. McKeeny. 1989. Sensitivity of Evapotranspiration in a Wheat Field, a Forest, and a Grassland to Changes in Climate and Direct Effects of Carbon Dioxide. *Climate Change* 14: 117–151.

Maugh, T. H., II. 1979. SO_2 Pollution May be Good for Plants. *Science* 205: 383.

Mauney, J. R., K. E. Fry, and G. Guinn. 1978. Relationship of Photosynthesis Rate to Growth and Fruiting of Cotton, Soybean, and Sunflowers. *Crop Sci.* 18: 259–263.

Mayer, A., and J. Mayer. 1974. Agriculture, the Island Empire. *Daedalus* Science and Its Public: The Changing Relationship. *J. Am. Acad. Arts Sci.* Summer: 84–95.

McNeill, W. H. 1991. American Food Crops in the Old World. In: *Seeds of Change,* H. J. Viola and C. Margolis (Eds), pp. 43–59. Smithsonian Institution Press, Washington, D.C.

McQuigg, J. D. 1977. Climatic Constraints on Food Grain Production. In: *Proceedings of the World Food Conference of 1976,* June 27–July 1. Iowa State University, pp. 387–394. The Iowa State University Press, Ames.

Meadows, D. H., D. L. Meadows, J. Randers, and W. Behrens. 1972. *The Limits to Growth.* Report of the Club of Rome. Universe Books, New York.

Mearns, I. O., C. Rosenzweig, and R. Goldberg. 1992. Effect of Changes in Interannual Climate Variability on CERES-Wheat Yields: Sensitivity and 2 × CO_2 General Circulation Model Studies. *Agr. For. Meteorol.* 62: 159–189.

Michaels, P. J. 1990. The Greenhouse Effect and Global Change: Review and Reappraisal. *Int. J. Environ. Stud.* 6: 55–71.

Michaels, P. J. 1991. Global Pollution's Silver Lining. *New Sci.* No. 1796, November 23: 40–43.

Michaels, P. J. 1992. *Sound and Fury: The Science and Politics of Global Warming.* Cato Institute, Washington, D.C.

Michaels, P. J. 1993. Benign Greenhouse. In: *Research and Exploration,* Vol. 9(2), pp. 222–233. National Geographic Society. Washington, D.C.

Michaels, P. J., and D. E. Stooksbury. 1992. Global Warming: A Reduced Threat? *Bull. Am. Meteorol. Soc.* 73: 1563–1577.

Michaels, P. J. , P. C. Knappenberger, and D. A. Gay. 1993. Addendum to: Regional and Seasonal Analyses of Ground-Based and Satellite Sensed Temperatures: Where's the Warming. Presented at the 8th Conference on Applied Climatology. pp. 147–152. American Meteorological Society, Anaheim, CA.

Miller, P. R., and J. R. McBride. 1975. Effects of Air Pollutants on Forests. In: *Responses of Plants to Air Pollution.* J. B. Mudd and T.T. Kozlowski (Eds.), pp. 195–235. Academic Press, New York.

Miller, J. E., F. L. Booker, E. L. Fiscus, A. S. Heagle, W. A. Pursley, S. F. Vozzo, and W. W. Heck. 1994. Ultraviolet-B Radiation and Ozone Effects on Growth, Yield, and Photosynthesis of Soybean. *J. Environ. Qual.* 23: 83–91.

Mitchell, J. F. B., S. Manabe, V. Meleshko, and T. Tokioka. 1990. Equilibrium Climate Change—and its Implications for the Future. In: *Climate Change: The IPCC Scientific Assessment,* J. T. Houghton et al. (Eds.), pp. 137–104. Cambridge University Press, Cambridge, U.K.

Moe, R., and L. M. Mortensen. 1986. CO_2 Enrichment in Norway. In: *Carbon Dioxide Enrichment of Greenhouse Crops,* Vol. 1. H. Z. Enock, and B. A. Kimball (Eds.), pp. 59–73. CRC Press, Boca Raton, FL.

Monastersky, R. 1993. Plants and Soils May Worsen Global Warming. *Sci. News* 143(7): 100–101.

Monteith, J. L. 1981. Climate Variation and the Growth of Crops. *Q. J. Roy. Meteorol. Soc.* 197: 749–774.

Mooney, H. A., B. G. Drake, R. J. Luxmoore, W. C. Oechel, and L. F. Pitelka. 1991. Predicting Ecosystem Responses to Elevated CO_2 Concentration. *BioScience* 41(2): 96–104.

Mooney, H. A., G. W. Koch, and C. B. Field. 1994. Potential Impacts of Asymmetrical Day–Night Temperature Increase on Biotic Systems. In: *Asymetric Change of Daily Temperature Range.* G. Kukla, T. K. Karl, and M. C. Riches (Eds.), U.S. Department of Energy 025, Washington, D.C.

Morison, J. I. L. 1993. Response of Plants to CO_2 under Water Limited Conditions. *Vegetatio* 104/105: 193–209.

Morison, J. I. L., and R. M. Gifford. 1984. Plant Growth and Water Use with Limited Supply in High CO_2 Concentrations. II. Plant Dry Weight, Partitioning and Water Use Efficiency. *Aust. J. Plant. Physiol.* 11: 375–384.

Morrison, M. 1992. The Ozone Scare. *Insight* April 6: 7–13, 34.

Mortensen, L. M. 1987. Review: CO_2 Enrichment in Greenhouses. Crop Responses. *Scientia Horticulturae* 33:1–25.

Mousseau, M. 1993. Effects of Elevated CO_2 on Growth, Photosynthesis and Respiration of Sweet Chestnut (*Castanea sativa* Mill.) *Vegetatio* 104/105: 413–419.

Myers, N. 1983. *A Wealth of Wild Species.* Westview Press, Boulder, CO.

Nance, J. J. 1991. *What Goes Up: The Global Assault on Our Atmosphere.* William Morrow, New York.

National Academy of Sciences. 1976. *Climate and Food. Climatic Fluctuations and U.S. Agricultural Production.* National Academy of Sciences, Washington, D.C.

National Academy of Sciences. 1977. *The Potential Contributions of Research. World Food and Nutrition Study.* Report of the Steering Committee. National Research Council, Washington, D.C.

National Academy of Sciences. 1991. *Policy Implications of Greenhouse Warming.* Synthesis Panel. National Academy Press, Washington, D.C.

National Academy of Sciences. 1992. *Policy Implications of Greenhouse Warming. Mitigation, Adaptation and the Science Base.* National Academy Press, Washington, D.C.

National Energy Outlook Committee of the United States Energy Association. 1991. Fifth Annual Assessment, Washington, D. C.

National Research Council. 1972. *Genetic Vulnerability of Major Crops.* National Academy of Sciences, Washington, D. C.

National Research Council. 1981. *The Water Buffalo: New Prospects for an Underutilized Animal.* National Academy Press, Washington, D.C.

National Research Council. 1983. *Changing Climate. Report of the Carbon Dioxide Assessment Committee.* National Academy Press, Washington, D.C.

Nederhoff, E. M. 1990. Technical Aspects, Management and Control of CO_2 Enrichment in Greenhouses. *Acta Horticult.* 268: 127–138.

Nederhoff, E. M. , K. Buitelaar, and H. W. de Ruiter. 1991. *Effects of CO_2 Fruit Load, Harvest Frequency and Leaf Pruning on Leaf Conductance, Chlorosis and Production of Aubergine (Eggplant),* pp. 48–50. Annual Report 1991. Glasshouse Crops Research Station, Naaldwijk, The Netherlands.

Nederhoff, E. M. 1994. *Effects of CO_2 Concentration on Photosynthesis, Transpiration and Production of Greenhouse Fruit Vegetable Crops.* Glasshouse Crops Research Station, Naaldwijk, The Netherlands.

Nederhoff, E. M., and G. Vegter. 1994. Photosynthesis of Stands of Tomato, Cucumber and Sweet Pepper Measured in Greenhouses under Various CO_2 Concentrations. *Ann. Bot.* 73: 353–361.

Neue, H. U. 1993. Methane Emission from Rice Fields. *BioScience* 43(7): 466–474.

Newell, R. E., and T. G. Dopplick. 1979. Questions Concerning the Possible Influence of Anthropogenic CO_2 on Atmospheric Temperature. *J. Appl. Meterol.* 18: 822–825.

Newman, J. E. 1982. *Impacts of Rising Atmospheric Carbon Dioxide Levels on Agricultural Growing Seasons and Crop Water Efficiencies. Environmental and Social Consequences of a Possible CO_2 Induced Climate Change by an Increase in Atmospheric Carbon Dioxide,* Vol. 2, Part B. Carbon Dioxide Res. Div., U.S. Department of Energy, Washington, D.C.

Newman, J. E. 1989. The Direct and Indirect Impacts of Climate Change on Crop Production. In: *Proceedings for Agricultural Science Centennial,* pp. 101–129. Central Agricultural Exp. Sta., Steinkjer, Norway, August 9–11.

Newton, P. C. D., H. Clark, C. C. Bell, E. M. Glasgow, and B. D. Campbell. 1994. Effect of Elevated CO_2 and Simulated Changes in Temperature on the Species Composition and Growth Rates of Pasture Turves. *Ann. Bot.* 73: 52–59.

Nie, D., H. He, G. Mo, M. B. Kirkham, and E. T. Kanemasu. 1992. Canopy Photosynthesis and Evapotranspiration of Rangeland Plants under Doubled Carbon Dioxide in Closed Top Chambers. *Agr. For. Meteorol.* 61: 205–217.

Nijs, I., I. Impens, and T. Behaeghe. 1988. Effects of Rising Atmospheric Carbon Dioxide Concentration on Gas Exchange and Growth of Perennial Ryegrass. *Photosynthetica* 22: 44–50.

Nilsson, A. 1992. *Greenhouse Earth,* pp. 119–122. John Wiley, New York.

Nonhebel, S. 1993. Effects of Changes in Temperature and CO_2 Concentration on Simulated Spring Wheat Yields in the Netherlands. *Climate Change* 24(4): 311–329.

Nordhaus, W. D. 1994. Expert Opinion on Climate Change. *Am. Sci.* 82(1): 45–51.

Norby, R. J. 1987. Nodulation and Nitrogenase Activity in Nitrogen-Fixing Woody Plants Stimulated by CO_2 Enrichment in the Atmosphere. *Plant. Physiol.* 71: 77–82.

Norby, R. J., and E. G. O'Neill. 1989. Growth Dynamics and Water Use of Seedlings of *Quercus Alba* L. in CO_2 Enriched Atmospheres. *New Phytol.* 111: 491–500.

Norby, R. J., C. A. Gunderson, S. D. Wullschleger, E. G. O'Neill, and M. K. McCracken. 1992. Productivity and Compensatory Responses of Yellow-Poplar Trees in Elevated CO_2. *Nature (London)* 357: 322–324.

Norman, A. G. 1962. The Uniqueness of Plants. *Am. Sci.* 50: 436–449.

Nweke, F. I. 1987. Marketing and Export of Cocoyam and Its Potentials for Food Sufficiency and Future Recovery of Nigeria. Paper presented at a Workshop on Cocoyam, National Root Crops Research Institute. Umudike, August 16–17, 1987. International Institute for Tropical Agriculture, Ibadan, Nigeria.

Oechel, W. C., and B. R. Strain. 1985. Native Species Responses to Increased Atmospheric Carbon Dioxide Concentration. In: *Direct Effects of Increasing Carbon Dioxide on Vegetation Strain.* B. R. Strain, and J. Cure (Eds.), pp. 117–154, U.S. Department of Energy DOE/ER-0238, Washington, D.C.

Onwueme, I. C. 1987. Strategies for Increasing Cocoyams (*Colocasia xanthosoma)* in the Nigerian Food Basket. Paper presented at the First National Workshop on Cocoyams. Umudike, Nigeria.

Oram, P. A. 1985. Sensitivity of Agricultural Production to Climate Change. *Climate Change* 7: 129–152.

Oram, P. A. 1989. Views on The New Global Context for Agricultural Research: Implications for Policy. In: *Climatic and Food Security.* International Rice Research Institute, Manila; AAAS, Washington, D.C.

Osbrink, W. L. A. , J. T. Trumble, and R. E. Wagner. 1987. Host Suitability of *Phaseolus lunata for Trichoplusia (Lepidoptera: Noctuidae)* in Controlled Carbon Dioxide Atmospheres. *Environ. Entomol.* 16: 210–215.

Overdieck, D. 1993. Elevated CO_2 and the Mineral Content of Herbaceous and Woody Plants. *Vegetatio* 104/105: 403–416.

Overdieck, D., D. Bossemeyer, and H. Lieth. 1984. Longterm Effects of an Increased CO_2 Concentration Level on Terrestrial Plants in Model-Ecosystems. I. Phytomass Production and Competition of *Trifolium repens* L. and *Lolium perenne* L. *Progress in Biometeorology* 3: 344–352.

Overdieck, D., C. H. Reid, and B. R. Strain. 1988. The Effects of Preindustrial and Future CO_2 Concentrations on Growth, Dry Matter Production and the C/N Relationship in Plants of Low Nutrient Supply: *Vigna unguiculata* (cowpeas), *Abelmoschus esculentus* (Okra) and *Raphanus sativus* (Radish). *Angew. Bot.* 62: 119–134.

Owensby, C. E., P I. Coyne, and L. M. Auen. 1989. Rangeland–Plant Response to Elevated CO_2. II. Large Chamber System. Series 054. *Response of Vegetation to Carbon Dioxide.* Office of Energy Research, U.S. Department of Energy, Washington, D.C.

Owensby, C. E., P. I. Coyne, and I. M. Auen. 1993. Nitrogen and Phosphorus Dynamics of a Tallgrass Prairie Ecosystem Exposed to Elevated Carbon Dioxide. *Plant, Cell Environ.* 16: 843–850.

Owensby, C. E., P. I. Coyne, J. M. Ham, J. M. Auen, and A. K. Knapp. 1994. Biomass Production in a Tallgrass Prairie Ecosystem Exposed to Ambient and Elevated Levels of CO_2. *Ecol. Appl.* 3: 644–653.

Paez, A., H. Hellmers, and B. R. Strain. 1983. CO_2 Enrichment, Drought Stress and Growth of Alaska Pea Plants (*Pisum sativum) Plant Physiol.* 58: 161–165.

Palutikof, J. P., T. M. L. Wigley, and G. Farmer. 1984. The Impact of CO_2 Induced Climate Change on Crop–yields in England and Wales. *Prog. Biometeorol.* 3: 320–334.

Parry, M. L., T. R. Carter, and N. T. Konijn (Eds.). 1988a. *The Impact of Climate Variations on Agriculture. Vol. I: Assessments in Cool Temperate and Cold Regions.* The International Institute for Applied Systems Analysis, United Nations Environment Program. Kluwer Academic Publishers, Boston.

Parry, M. L., T. R. Carter, and N. T. Konijn (Eds.). 1988b. *The Impact of Climate Variations on Agriculture. Vol. 2: Assessments in Semi-Arid Regions.* International Institute for Applied Systems Analysis, United Nations Environment Program. Kluwer Academic Publishers, Boston.

Parry, M. R. 1990. *Climate Change and World Agriculture.* Earthscan Publications, London.

Patterson, D. T. 1982. Effects of Light and Temperature on Weed/Crop Growth and Competition. In: *Biometeorology in Integrated Pest Management.* J. L. Hatfield and I. J. Thompson (Eds.), pp. 407–420. Academic Press, New York.

Patterson, D. T. 1986. Responses of Soybean (Glycine Max) and Three C_4 Grass Weeds to CO_2 Enrichment During Drought. *Weed Sci.* 34: 203–210.

Patterson, D. T., and E. P. Flint. 1979. Effects of Simulated Field Temperatures and Chilling in Itchgrass (*Rottboellia exaltata,* Corn (*Zea mays),* and Soybean (*Glycine max). Weed Sci.* 27: 645–650.

Patterson, D. T., and E. P. Flint. 1980. Potential Effects of Global Atmospheric Carbon Dioxide Enrichment on the Growth and Competitiveness of C_3 and C_4 Weed and Crop Plants. *Weed Sci.* 28: 71–75.

Patterson, D. T., E. P. Flint, and J. L. Beyers. 1984. Effects of CO_2 Enrichment on Competition Between a C_4 Weed and C_3 Crop. *Weed Sci.* 32: 101–105.

Patterson, D. T., and E. P. Flint. 1990. Implications of Increasing Carbon Dioxide and Climate Change for Plant Communities and Competition in Natural and Managed Ecosystems. In: *Impact of Carbon Dioxide, Trace Gases, and Climate Change on Global Agriculture.* B. A. Kimball et al. (Eds.), pp. 83–110. ASA Special Publication No. 53. American Society of Agronomy Madison, WI.

Pearcy, R. W., and O. Bjorkman. 1983. Physiological Effects. In: *CO_2 and Plants, The Response of Plants to Rising Levels of Atmospheric Carbon Dioxide,* E. R. Lemon (Ed), pp. 65–105. AAAS Symposium 84. Westview Press, Boulder, CO.

Peart, R. M., J. W. Jones, R. B. Curry, K. J. Boote, and L. H. Allen, Jr. 1989. Impact of Climate Change on Crop Yield in the Southeastern USA: A Simulation Study. In: *The Potential Effects of Global Climate Change in the United States,* Appendix C, *Agriculture,* Vol. 1, J. Smith, and D. Tirpack (Eds.), pp. 2.1–2.54. U.S. Environmental Protection Agency, Washington, D.C.

Pendleton, D. F., and G. M. van Dyne. 1982. *Research Issues in Grazinglands Under Changing Climates.* U.S. Department of Energy, Office of Energy Research, Carbon Dioxide Research Divison 013, Vol. 2. Part 16. Washington, D.C.

Pendleton, J. W., and T. L. Lawson. 1989. Climatic Variability and Sustainability of Crop Yields in the Humid Tropics. In: *Climate and Food Security*, pp. 57–68. International Symposium on Climate Variability and Food Security in Developing Countries, 5–9 February, 1987. New Delhi, India. International Rice Research Institute, Manila, and American Association for the Advancement of Science, Washington, D.C.

Penning de Vries, F. W. T., H. Van Keulen, C. A. Van Diepen, I. G. A. M. Noy, and J. Goudriaan. 1989. Simulated Yields of Wheat and Rice in Current Weather and When Ambient CO_2 Has Doubled. In: *Climate and Food Security*, pp. 347–357. International Rice Research Institute, Manila and American Association for the Advancement of Science, Washington, D.C.

Penning de Vries, F. W. T. 1992. Rice Production and Climate Change. In: *Systems Approaches for Agricultural Development*. F. W. T. Penning de Vries, P. S. Teng, and K. Metselaar (Eds.). Kluwer Academic, Dordrecht, The Netherlands.

Per Pinstrup-Andersen. 1994. *World Food Trends and Future Food Security*. Food Policy Report, The International Food Policy Research Institute, Washington, D.C.

Peters, R. L., and J. D. S. Darling. 1985. The Greenhouse Effect and Nature Resources. *BioScience* 35(11): 707–717.

Petranek, S. 1993. The Force of Nature. *Life*. September: 30–37.

Pettersson, R., H. S. J. Lee, and P. G. Jarvis. 1993. The Effect of CO_2 Concentration on Barley. *Vegetatio* 104/105: 462–463.

Pimentel, D. 1991. Global Warming, Population Growth, and Natural Resources for Food Production. *Soc. Nat. Resourc.* 4: 347–363.

Pimentel, D., N. Brown, F. Vecchio, V. La Capra, H. Hausman, O. Lee, A. Diaz, J. Williams, S. Cooper, and E. Newburger. 1992. Ethical Issues Concerning Potential Global Climate Change on Food Production. *J. Agr. Environ. Ethics*, 5(2): 113–146.

Pimm, S. L., and A. M. Sugden. 1994. Tropical Diversity and Global Change. *Science* 263: 933–934.

Polley, H. W., H. B. Johnson, B. D. Marino, and H. S. Mayeux. 1993. Increase in C_3 Plant Water-Use Efficiency and Biomass Over Glacial to Present CO_2 Concentrations. *Nature (London)* 361: 61–64.

Pond, W. G., R. A. Merkel, L. D. McGilliard, and V. J. Rhodes (Eds). 1980. Animal Agriculture. *Research to Meet Human Needs in the 21st Century*. Westview Press, Boulder, CO.

Poorter, H. 1993. Interspecific Variation in the Growth Response of Plants to an Elevated Ambient CO_2 Concentration. *Vegetatio* 104: 77–97.

Porter, M. A., and B. Grodzinski. 1985. CO_2 Enrichment of Protected Crops. *Hort. Rev.* 7: 345–398.

Postel, S. 1989. *Water for Agriculture: Facing the Limits*. Worldwatch Paper 93. Worldwatch Institute, Washington, D. C.

Postel, S. 1993. *Facing Water Scarcity. State of the World*. pp. 22–41. W. A. Norton, New York.

Prior, S. A., H. H. Rogers, G.B. Runion, and J. R. Mauney. 1994. Effects of Free–Air Enrichment on Cotton Root Growth. *Agr. For. Meteorol.* 70: 69–86

Rabinowitch, E. I. 1956. *Photosynthesis and Related Processes*, Vol. 2, Part 2. Interscience Publishers, New York.

Ramanathan, V. 1988. The Greenhouse Theory of Climate Change: A Test by an Inadvertent Global Experiment. *Science* 240: 293–299.

Rao, N. G. P., G. R. K. Rao, and H. S. Acharya. 1989. Yield Stability of Sorghum and Millet Across Climates. In: *Climate and Food Security.* pp. 165–186. International Rice Research Institute, Manila and American Association for the Advancement of Science, Washington, D.C.

Raschke, K. 1986. The Influence of the CO_2 Content of the Ambient Air on Stomatal Conductance and the CO_2 Concentration in Leaves. In: *Carbon Dioxide Enrichment of Greenhouse Crops,* Vol. 2, H. Z. Enoch, and B. A. Kimball (Eds.), pp. 87–102. CRC Press, Boca Raton, FL.

Raynaud, D., J. Jouzel, J. M. Barnola, J. Chappellaz, R. J. Delmas, and C. Lorius. 1993. The Ice Record of Greenhouse Gases. *Science* 259: 926–934.

Reardon, J. C., J. R. Lambert, and B. Acock. 1990. *The Influence of Carbon Dioxide Enrichment on the Seasonal Patterns of Nitrogen Fixation in Soybeans.* Series 016. *Response of Vegetation to Carbon Dioxide.* Office of Energy Research, U.S. Department of Energy, Washington, D.C.

Reddy, K. R., and W. H. Smith (Eds.). 1987. Aquatic Plants for Water Treatment and Resource Recovery. Magnolia Publ., Orlando, FL.

Reddy, V. R., B. Acock, and M. C. Acock. 1989. Seasonal Carbon and Nitrogen Accumulation in Relation to Net Carbon Dioxide Exchange in a Carbon Dioxide-Enriched Soybean Canopy. *Agron. J.* 81: 78–83.

Reid, W. V. 1994. The Economic Realities of Biodiversity. *Issues Sci. Technol.* 10(2): 48–55.

Reifsnyder, W. E. 1989. A Tale of Ten Fallacies: The Skeptical Enquirer's View of the Carbon Dioxide/Climate Controversy. *Agr. For. Meteorol.* 47: 349–371.

Reuveni, J., J. Gale, and A. M. Mayer. 1993. Reduction of Respiration by High Ambient CO_2 and the Resulting Error in Measurements of Respiration made with O_2 Electrodes. *Ann. Bot.* 72: 129–131.

Revelle, R. 1982. Carbon Dioxide and World Climate. *Sci. Am.* 247(2): 35–43.

Revelle, R., and H. E. Suess. 1957. Carbon Dioxide Exchange Between Atmosphere and Ocean and the Question of an Increase of Atmospheric CO_2 During the Past Decades. *Tellus* 9: 18–27.

Rich, S. 1963. The Role of Stomata in Plant Disease. Bull. Connecticut Agr. Exp. Sta. 664. New Haven.

Riebsame, W. E. 1989. *Assessing the Social Implications of Climate Fluctuation: A Guide to Climate Impact Studies.* World Climate Impacts Programme, United Nations Environmental Program, Nairobi, Kenya.

Ritchie, J. T., B. D. Baer, and T. Y. Chou. 1989. Effect of Global Climate Change on Agriculture, Great Lakes Region. In: *The Potential Effects of Global Climate Change on the United States.* Appendix C. *Agriculture,* Vol. 1, pp. 1.1–1:43. U.S. Environmental Protection Agency, Washington, D.C.

Robertson, G. P. 1993. Fluxes of Nitrous Oxide and Other Nitrogen Trace Gases from Intensively Managed Landscapes: A Global Perspective. In: *Agricultural Ecosystem Effects on Trace Gases and Global Climate Change.* pp. 95–108. ASA Special Publication No. 55. American Society of Agronomy, Madison, WI.

Rogers, H. H., and R. C. Dahlman. 1993. Crop Responses to CO_2 Enrichment. *Vegetatio* 104/105: 117–131.

Rogers, H. H., G. E. Bingham, J. F. Thomas, J. M. Smith, D. W. Israel, and K. A. Surano. 1981. Effects of Long-Term CO_2 Concentrations on Field-Grown Crops and Trees. In: *Global Dynamics of Biospheric Carbon,* S. Brown (Ed.), pp. 9–45. U.S. Department of Energy, Washington, D.C.

Rogers, H. H., G. E. Bingham, J. D. Cure, J. M. Smith, and K. A. Surano. 1983a. Responses of Selected Plant Species to Elevated Carbon Dioxide in the Field. *J. Environ. Qual.* 12(14): 569–574.

Rogers, H. H., J. F. Thomas, and G. E. Bingham. 1983b. Response of Agronomic and Forest Species to Elevated Atmospheric Carbon Dioxide. *Science* 220: 428–429.

Rogers, H. H., N. Sionit, J. D. Cure, J. M. Smith, and G. E. Bingham. 1984. Influence of Elevated Carbon Dioxide on Water Relations in Soybeans. *Plant Physiol.* 74: 233–238.

Rogers, H. H., J. D. Cure, and J. M. Smith. 1986. Soybean Growth and Yield Response to Elevated Carbon Dioxide. *Agr., Ecosyst. Environ.* 16: 113–128.

Rogers, H. H., C. M. Peterson, J. N. McCrimmon, and J. D. Cure. 1992a. Response of Plant Roots to Elevated Atmospheric Carbon Dioxide. *Plant, Cell Environ.* 15: 749–752.

Rogers, H. H., S. A. Prior, and E. G. O'Neill. 1992b. Cotton Root and Rhizosphere Responses to Free-Air CO_2 Enrichment. *Crit. Rev. Plant Sci.* 11(2–3): 251–263.

Rogers, H. H., L. H. Allen, Jr., B. A. Kimball, S. B. Idso, J. E. Miller, S. L. Rawlins, and R. C. Dahlman. 1992c. Direct Effects of Rising CO_2 on Crops. Public Hearing of the Committee of Enquiry "Protecting the Earth's Atmosphere," 17–18 February 1992. Deutscher Bundestag, Bundeshouse Bonn, Germany.

Rogers, H. H., and G. B. Runion. 1994. Plant Responses to Atmospheric CO_2 Enrichment with Emphasis on Roots and the Rhizosphere. *Environ. Poll.* 83: 155–189.

Rogers, H. H., G. B. Runion, S. V. Krupa, and S. A. Prior. 1995. Plant Responses to Enrichment: Implications in Root-Soil-Microbe Interactions. In: *Advances in CO_2 Effects Research.* L. H. Allen, Jr. et al. (Eds.). ASA Special Publication (in press).

Rose, Elise. 1989. Direct (Physiological) Effects of Increasing CO_2 on Crop Plants and Their Interactions With Indirect (Climatic) Effects. In: *The Potential Effects of Global Climate Change on the United States,* Appendix C. *Agriculture,* Vol. 2. pp. 7.1–7.37. U.S. Environmental Protection Agency, Washington, D.C.

Rosenberg, N. J. (Ed.). 1980. *Drought in the Great Plains: Research on Impacts and Strategies.* Water Resources Publications, Littleton, CO.

Rosenberg, N. J. 1981. The Increasing CO_2 Concentration in the Atmosphere and Its Implications on Agricultural Productivity. I. Effects on Photosynthesis, Transpiration and Water Use Effeciency. *Climate Change* 3: 265–279.

Rosenberg, N. J. 1982. The Increasing CO_2 Concentration in the Atmosphere and Its Implication on Agricultural Productivity. II. Effects Through CO_2-Induced Climate Change. *Climate Change* 4: 239–254.

Rosenberg, N. J. 1983. *A Synthesis and Some Questions on the Possible Effects of Increased Atmospheric CO_2 Concentration on Agriculture.* Agronomy Abstracts. p. 16. American Society of Agronomy, Madison, WI.

Rosenberg, N. J. 1988. *Climate Change. A Primer.* Resources for the Future, Washington, D.C.

Rosenberg, N. J. 1989. Potential Effects on Crop Production of Carbon Dioxide Enrichment in the Atmosphere and Greenhouse-Induced Climate Change. In: *Climate and Food Security.* pp. 359– 373. Resources for the Future, International Rice Research Institute, Manila and American Society for the Advancement of Sciences, Washington, D.C.

Rosenberg, N. J. (Ed). 1992a. A Methodology for Assessing Regional Agricultural Conseqences of Climate Change: Application to the Missouri-Iowa-Nebraska-Kansas (MINK) Region. Contributions from the Study Processes for Identifying Regional Influences of and Responses to Increasing Atmospheric CO_2 and Climate Change—The MINK Project. *Agr. For. Meteorol.* 59(1–2): 1–126.

Rosenberg, N. J. 1992b. Adaptation of Agriculture to Climate Change. *Climate Change* 21: 385–405.

Rosenberg, N. J. 1993. Towards an Integrated Impact Assessment of Climate Change: The MINK Study. Special Issue of *Climate Change* 24 (1 and 2), pp. 1–173.

Rosenberg, N. J., and P. R. Crosson. 1991. The MINK Project: A New Methodology for Identifying Regional Influences of, and Responses to, Increasing Atmospheric CO_2 and Climate Change. *Environ. Conserv.* 18(4): 313–322.

Rosenberg, N. J., M. S. McKenney, and P. Martin. 1989. Evapotranspiration in a Greenhouse Warmed World: A Review and a Simulation. *Agr. For. Meteorol.* 47: 303–320.

Rosenberg, N. J., B. A. Kimball, P. Martin, and C. F. Cooper. 1990. From Climate and CO_2 Enrichment to Evapotranspiration. In: *Climate Change and U. S. Water Resources,* P. E. Waggoner (Ed.), pp. 151–175. John Wiley, New York.

Rosenzweig, C. 1989. Potential Effects of Climate Change on Agricultural Production in the Great Plains. A Simulation Study. In: *The Potential Effects of Global Climate Change in the United States.* Appendix C. *Agriculture,* Vol. 1, pp. 3.1–3.43. U.S. Environmental Protection Agency, Washington, D.C.

Rosenzweig, C., and D. Hillel. 1993a. The Dust Bowl of the 1930's. Analog of Greenhouse Effect in the Great Plains. *J. Environ. Qual.* 22: 9–22.

Rosenzweig C., and D. Hillel. 1993b. Agriculture in a Greenhouse World. In: *Research and Exploration,* Vol. 9(2), pp. 208–221. National Geographic Society, Washington, D.C.

Rosenzweig, C., and M. L. Parry. 1994. Potential Impact of Climate Change on World Food Supply. *Nature (London)* 367: 133–138.

Rosenzweig, C., M. L. Parry, G. Fischer, and K. Frohberg. 1993. *Climate Change and World Food Supply.* Research Report No. 3, Environmental Change Unit, Uiversity of Oxford, Oxford, U.K.

Rozema, J. 1993. Plant Responses to Atmospheric Carbon Dioxide Enrichment: Interactions with Some Soil and Atmospheric Conditions. *Vegetatio* 104/105: 173–190.

Rozema, J., H. Lambers, S. C. van de Geijn, and M. L. Cambridge (Eds.), 1993. *CO_2 and Biosphere.* Kluwer Academic Publishers, Boston.

Runion, G. B., E. A. Curl, H. H. Rogers, P. A. Backman, R. Rodriquez–Kabana, and B. E. Helms. 1994. Effects of Free–Air CO_2 Enrichment on Microbial Populations in the Rhizosphere and Phyllosphere of Cotton. *Agr. For. Meteor.* 70: 117–130.

Ruttan, V. E. (Ed). 1990. *Resource and Environmental Constraints on Sustainable Growth in Agricultural Production: Report of a Dialogue.* Staff Paper P90-33. University of Minnesota Department of Agricultural and Applied Economics, St. Paul, MN.

Sasek, T. W., and B. R. Strain, 1990. Implications of Atmospheric CO_2 Enrichment and Climate Change for the Geographical Distribution of Two Introduced Vines in the USA. *Climate Change* 16: 31–51.

Satake T., and S. Yoshida. 1978. High Temperature Induced Sterility in Indica Rices at Flowering. *Jpn. J. Crop Sci.* 47: 6–17.

Saunders, D. A. (Ed.). 1991. What for the Nontraditional Warm Areas. Proc. of the International Conference July 23–Aug. 3, 1990. Foz do Iguaeu, Brazil. United Nations Development Program. International Maize and Wheat Improvement Centre, Mexico.

Schelling, T. C. 1983. Climate Change: Implications for Welfare and Policy. In: *Changing Climate.* pp. 449–482. National Research Council, National Academy Press, Washington, D.C.

Schlesinger, M. E. 1993. Greenhouse Policy. In: *Research and Exploration,* Vol. 9(2), pp. 159–172. National Geographic Society, Washington, D.C.

Schlesinger, W. H. 1993. Response of the Terrestrial Biosphere to Global Climate Change and Human Perturbation. *Vegetatio* 104/105: 295–305.

Schmidtmann, E. T., and J. A. Miller. 1989. Effect of Climatic Warming on Population of the Horn Fly, With Associated Impact on Weight Gain and Milk Production in Cattle. In: *The Potential Effects of Global Climate Change on the United States.* Appendix C. *Agriculture.* Vol. 2, pp. 12.1–12.11. U.S. Environmental Protection Agency, Washington, D.C.

Schneider, S. H. 1989. The Changing Climate. *Sci. Am.* 261(3): 70–79.
Schneider, S. H. 1990. The Global Warming Debate Heats Up: An Analysis and Perspective. *Bull. Am. Meteorol. Soc.* 71: 1292–1304.
Schneider, S. H. 1992. Will Sea Levels Rise or Fall? *Nature (London)* 356: 11–12.
Schneider, S. H. 1993. Degrees of Certainty In: *Research and Exploration,* Vol. 9(2), pp. 173–190. National Geographic Society, Washington, D.C.
Schneider, S. H. 1994. Detecting Climate Change Signals: Are There Any "Fingerprints"? *Science* 263: 341–347.
Schuh, G. E. 1989. Agricultural Policies for Climate Changes Induced by Greenhouse Gases. In: *The Potential Effects of Global Climate Changes on the United States.* Appendix C. *Agriculture.* Vol. 2, pp. 13.1–13.14. U.S. Environmental Protection Agency, Washington, D.C.
Schultz, T. W. 1982. *The Dynamics of Soil Erosion in the United States: A Critical Review.* Agricultural Economics Paper No. 82.8. University of Chicago. Conference on Soil Conservation, Agricultural Council of America, March 17, 1982, Washington, D.C.
Sedjo, R. A., and A. M. Solomon. 1989. Climate and Forests. In: *Greenhouse Warming: Abatement and Adaptation.* N. S. Rosenberg, W. E. Easterling III, P. R. Crosson, and J. Darmstadter (Eds.). Resources for the Future, Washington, D.C.
Seitz, F., K. Bendetsen, R. Jastrow, and W. A. Nierenberg. 1989. *Scientific Perspectives on the Greenhouse Problem.* George C. Marshall Institute, Washington, D.C.
Seshu, D. V., T. Woodhead, D. P. Garrity, and L. R. Oldeman. 1989. Effect of Weather and Climate on Production and Vulnerability of Rice. In: *Climate and Food Security,* pp. 93–121. International Symposium on Climate Variability and Food Security in Developing Countries, 5–9 February, 1987, New Delhi, India. International Rice Research Institute, Manila and American Association for the Advancement of Science, Washington, D.C.
Shaver, G. R., W. D. Billings, E. S. Chapin III, A. E. Giblin, K. J. Nadelhoffer, W. C. Oechel, and E. B. Rostetter. 1992. Global Change and the Carbon Balance of Arctic Ecosystems. *BioScience* 42(6): 433–441.
Shu G., and C. W. Cady (Eds.). 1991. *Climate Variations and Change: Implications for Agriculture in the Pacific Rim.* University of California, Davis.
Singer, S. F., R. Revelle, and C. Starr. 1991. What to Do about Greenhouse Warming: Look Before You Leap. *Cosmos* 1(1): 28–33.
Sionit, N., H. Hellmers, and B. R. Strain. 1980. Growth and Yield of Wheat under Carbon Dioxide Enrichment and Water Stress Conditions. *Crop Sci.* 20: 687–690.
Sionit, N., B. R. Strain, H. Hellmers, and P. J. Kramer. 1981. Effects of Atmospheric Carbon Dioxide Concentration and Water Stress on Water Relations of Wheat. *Bot. Gazette* 142: 191–196.
Sionit, N., and P. J. Kramer, 1986. Woody Plant Reactions to CO_2 Enrichment. In: *Carbon Dioxide Enrichment of Greenhouse Crops,* H. Z. Enoch and B. A. Kimball (Eds.), pp. 69–85. CRC Press, Boca Raton, FL.
Sionit, N., and D. T. Petterson. 1986. Responses of C_4 Grasses to Atmospheric CO_2 Enrichment. II. Effect of Water Stress. *Crop Sci.* 25: 533–537.
Sivakumar, M. V. K. 1989. Climate Vulnerability of Sorghum and Millet. In: *Climate and Food Security.* pp. 187–192. International Rice Research Institute, Manila and American Association for the Advancement of Science, Washington, D.C.
Slack, G. 1986. CO_2 Enrichment of Tomato Crops. In: *Carbon Dioxide Enrichment of Greenhouse Crops.* Vol. 2. H. Z. Enoch and B. A. Kimball (Eds.), pp. 151–163. CRC Press, Boca Raton, FL.
Smajstria, A. G. 1993. Microirrigation for Citrus in Florida. *HortScience* 28(4): 295–298.
Smit, B., L. Ludlow, and M. Brklacich. 1988. Implications of a Global Climatic Warming for Agriculture: A Review and Appraisal. *J. Environ. Qual.* 17: 519–527.

Smith, J. B., and D. A. Tirpak (Eds.). 1989a. *The Potential Effects of Global Climate Change on the United States.* Appendix C. *Agriculture,* Vol. 1. EPA Office of Policy, Planning and Evolution, Washington, D.C.

Smith, J. B., and D. A. Tirpak (Eds.). 1989b. *The Potential Effects of Global Climate Change in the United States.* Appendix C. *Agriculture,* Vol. 2. EPA Office of Policy, Planning and Evolution, Washington, D.C.

Smith, J. B., and D. A. Tirpak (Eds.). 1989c. *The Potential Effects of Global Climate Change on the United States.* Report to Congress. EPA Office of Policy, Planning and Evaluation, Washington, D.C.

Solomon, A. M., and D. C. West. 1985. Potential Responses of Forests to CO_2-Induced Climate Change. In: *Characterization of Information Requirements for Studies of CO_2 Effects: Water Resources, Agriculture, Fisheries, Forests and Human Health.* M. R. White (Ed.), pp. 147–169. U.S. Department of Energy, DOE/ER-0236, Washington, D.C.

Stem, E., G. A. Mertz, J. D. Stryker, and M. Huppi. 1989. Changing Animal Disease Patterns Induced by the Greenhouse Effect. In: *The Potential Effects of Global Climate Change on the United States.* Appendix C. *Agriculture,* Vol. 2, pp. 11.1–11.37. U.S. Environmental Protection Agency, Washington, D.C.

Stinner, B. R., R. A. J. Taylor, R. B. Hammond, F. F. Purrington, D. A. McCartney, N. Rodenhouse, and G. W. Barrett. 1989. Potential Effects of Climate Change on Plant–Pest Interactions. In: *The Potential Effects of Global Climate Change on the United States.* Appendix C, *Agriculture,* Vol. 2, pp. 8.1–8.35. U.S. Environmental Protection Agency, Washington, D.C.

Stitt, M. 1991. Rising CO_2 Levels and Their Potential Significance for Carbon Flow in Photosynthetic Cells. *Plant Cell Environ.* 14: 741–762.

Stolarski, R. , R. Bojkow, L. Bishop, C. Zerefos, J. Staehelin, and J. Zawodny. 1992. Measured Trends in Stratospheric Ozone. *Science* 256: 342–349.

Strain, B. R. 1985. Background on the Response of Vegetation to Atmospheric Carbon Dioxide Enrichment. In: *Direct Effects of Increasing Carbon Dioxide on Vegetation.* pp. 1–10. U.S. Department of Energy, DOE/ER-0238. Washington, D.C.

Strain, B. R. 1987. Direct Effect of Increasing Atmospheric CO_2 on Plants and Ecosystems. *Trends Ecol. Evol.* 2: 18–21.

Strain, B. R. 1992. Atmospheric Carbon Dioxide: A Plant Fertilizer? *New Biol.* 4(2): 87–89.

Strain, B. R., and F. F. Bazzaz. 1983. Terrestrial Plant Communities. In: *CO_2 and Plants. The Response of Plants to Rising Levels of Atmospheric Carbon Dioxide.* pp. 177–222. AAA's Selected Symposium 84. Westview Press, Boulder, CO.

Strain, B. R., and J. D. Cure (Eds.). 1985. Direct Effects of Increasing Carbon Dioxide on Vegetation. U.S. Department of Energy, DOE/ER-0238, Washington, D.C.

Strain, B. R., and J. D. Cure (Eds.). 1994. Direct Effects of Atmospheric CO_2 Enrichment on Plants and Ecosystems. ORNL/CDIAC-70. Carbon Dioxide Information Analysis Center, Oak Ridge National Laboratory, TN.

Strain, B. R., and J. D. Cure. 1986. Direct Effects of Atmospheric CO_2 Enrichment on Plants and Ecosystems: A Bibliography with Abstracts. Oak Ridge National Laboratory, ORNL/CDIC-13, Oak Ridge, TN.

Strommen, N. D. 1992. Climate and Crop Yield. In: *Research and Exploration,* Vol. 8(1), pp. 10–21. National Geographic Society, Washington, D.C.

Stulen, I., and J. den Hertog. 1993. Root Growth and Functioning under Atmospheric CO_2 Enrichment. *Vegetatio* 104/105: 99–115.

Sullivan, J. H. 1994. Temporal and Fluency Responses of Tree Foliage to UV-B Radiation. In: *Proceedings of the NATO Advances Research Conference on Stratospheric Ozone Depletion/UV-B Radiation in the Environment.* R. H. Biggs (Ed.). Springer-Verlag, Berlin (in press).

Sullivan, J. H., and A. H. Teramura. 1988. Effects of Ultra-Violet-B Irradiation on Seedling Growth in the Pinaceae. *Am. J. Bot.* 75: 225–230.

Sullivan, J. H., and A. H. Teramura. 1990. Field Study of the Interaction Between Solar Ultraviolet-B Radiation and Drought on Photosynthesis and Growth in Soybean. *Plant Physiol.* 92: 141–146.

Sullivan, J. H., and A. H. Teramura. 1992. The Effects of Ultraviolet-B Radiation on Loblolly Pine. *Trees* 6: 115–120.

Sullivan, J. H., and A. H. Teramura. 1994. The Effect of UV-B Radiation on Loblolly Pine. III. Interaction with CO_2 Enhancement. *Plant, Cell Environ.* 17(3): 311–317.

Sullivan, J. M., et al. 1995. Comparison of the Response of Soyean to Supplemental UV-B Radiation Supplied by Either Square-Wave or Modulated Sstems. In: *Proceedings of a NATO Advanced Research Conference in Stratospheric Ozone Depletion/UV-B Radiation in the Environment.* R. H. Biggs (Ed.). NATO, Brussels.

Surano, K. A., and J. H. Shinn. 1984. *CO_2 and Water Use Effects on Yield, Water–Use Efficiency and Photosynthate Partioning in Field Grown Corn.* Lawrence Livermore National Laboratory, UCLL–90771, Livermore, CA.

Swennen, R. 1990. *Plantain Cultivation under West African Conditions.* International Institute of Tropical Agriculture, Ibadan, Nigeria.

Taylor, K. E., and M. C. MacCracken. 1990. Projected Effects of Increasing Concentrations of Carbon Dioxide and Trace Gases on Climate. In: *Impact of Carbon Dioxide, Trace Gases, and Climate Change on Global Agriculture.* pp. 1–17. ASA Publication No. 53. American Society of Agronomy, Madison, WI.

Teramura, A. H. 1983. Effects of Ultraviolet-B Radiation on the Growth and Yield of Crop Plants. *Physiol. Plant.* 58: 415–427.

Teramura, A. H. 1987. Ozone Depletion and Plants. In: *Assessing the Risks of Trace Gases That Can Modify the Stratosphere,* Vol 8. J. S. Hoffman (Ed), pp. 1–75. U.S. Environmental Protection Agency, Washington, D.C.

Teramura, A. H. 1990. Implications of Stratospheric Ozone Depletion Upon Plant Production. *HortScience* 25(12): 1557–1560.

Teramura, A. H., and J. H. Sullivan. 1987. Soybean Growth Responses to Enhanced Levels of Ultraviolet-B Radiation under Greenhouse Conditions. *Am. J. Bot.* 74: 975–979.

Teramura, A. H., and J. H. Sullivan. 1988. Effects of Ultraviolet-B Radiation on Soybean Yield and Seed Quality: A Six Year Field Study. *Environ. Poll.* 53: 466–468. (See also *Physiol. Plant.* 80: 5–11).

Teramura, A. H., R. H. Biggs, and S. V. Kossuth. 1980. Effect of Ultraviolet-B Irradiances on Soybeans. II. Interaction Between Ultraviolet-B and Photosynthetically Active Radiation on Net Photosynthesis, Dark Respiration and Transpiration. *Plant Physiol.* 65: 483–488.

Teramura, A. H., J. H. Sullivan, and L. H. Ziska. 1990. Interaction of Elevated Ultraviolet-B Radiation and CO_2 on Productivity and Photosynthetic Characteristics in Wheat, Rice, and Soybean. *Plant Physiol.* 94: 470–475.

Thompson, G. B., and B. G. Drake. 1994. Insects and Fungi on a Sedge and a C_3 Grass Exposed to Elevated Atmospheric CO_2 Concentrations in Open-Top Chambers in the Field. *Plant, Cell Environ.* 17: 1161–1167.

Thompson, L. M. 1975. Weather Variability, Climate Change and Grain Production. *Science* 188: 535–541.

Thompson, L. M. 1980. Climate Change and World Grain Production. In: *The Politics of Food.* D. Gale Johnson (Ed.), pp. 100–123. The Chicago Council of Foreign Relations, Chicago, IL.

Tissue, D. T., and W. C. Oechel. 1987. Response of *Eriophorum vaginatum* to Elevated CO_2 and Temperature in the Alaskan Tussock Tundra. *Ecology* 68: 401–410.

Titus, J. G. 1990a. Effect of Climate Change on Sea Level Rise and the Implications for World Agriculture. *HortScience* 25(12): 1567–1572.

Titus, J. G. 1990b. Strategies for Adapting to the Greenhouse Effect. *APA J.*, Summer: 311–323.

Titus, J. G., C. Y. Kuo, M. J. Gibbs, T. B. LaRoche, M. K. Webb, and J. O. Waddell. 1993. Greenhouse Effect, Sea Level Rise and Coastal Drainage Systems. *J. Water Resour. Plan. Manage.* 113(2): 216–227.

Tognoni, F., A. H. Halevy, and S. H. Wittwer. 1967. Growth of Bean and Tomato Plants as Affected by Root Absorbed Growth Sustances and Atmospheric Carbon Dioxide. *Planta* 72: 43–52.

Tolbert, N. E., and I. Zelitch. 1983. Carbon Metabolism. In: *CO_2 and Plants,* E. R. Lemon, (Ed.), pp. 21–64. AAAS Selected Symposium 84. Westview Press, Boulder, CO.

Tripp, K. E., W. K. Kroen, M. M. Peet, and D. H. Willits. 1992. Fewer Whiteflies Found on CO_2 Enriched Greenhouse Tomatoes with High C:N Ratios. *HortScience* 27: 1079–1080.

UNEP, 1985. An Assessment of the Role of Carbon Dioxide and Other Greenhouse Gases in Climate Variations and Associated Impacts. United Nations Environmental Program/World Meteorological Organization/International Council of Scientific Unions. The Villach Conference. WMO. Geneva.

U.S. Department of Agriculture. 1972. *Extent and Cost of Weed Control with Herbicides, and an Evaluation of Important Weeds, 1968.* USDA-ARS. H 1, Washington, D.C.

U.S. Department of Energy. 1989. *Energy and Climate Change.* Report of the DOE Multi-Laboratory Climate Change Committee, December, U.S. Department of Energy, Washington, D.C.

U.S. Department of Energy. 1992. *Modeling the Response of Plants and Ecosystems to Elevated CO_2 and Climate Change.* Report TRO 54. U. S. Department of Energy, Washington, D.C.

U.S. Environmental Protection Agency. 1977. *Complilation of Air Pollutant Emission Factors.* AP-42 USEPA. Office of Air and Waste Management, Research Triangle Park, N.C.

Urban, D. L., and H. H. Shugart. 1989. Forest Response to Climate Change: A Simulation Study for Southeastern Forests. In: *The Potential Effects of Global Climate Change on the United States* (EPA-230-05-89-054). Appendix D. J. B. Smith and D. A. Tirpak (Eds.). U. S. Environmental Protection Agency, Washington, D.C.

Van Berkel, N. 1986. CO_2 Enrichment in the Netherlands. In: *Carbon Dioxide Enrichment of Greenhouse Crops.* Vol. I. H. Z. Enoch and B. A. Kimball (Eds.), pp. 17–33. CRC Press, Boca Raton, FL.

Van de Staaij, J. W. M., G. M. Lenssen, M. Stroetenga, and J. Rozema. 1993. The Combined Effects of Elevated CO_2 Levels and UV-B Radiation on Growth Characteristics of *Elymus athericus (E. pycnanathus). Vegetatio* 104/105: 433–439.

Van Helmont, J. B. 1648. *Ortus Medicinae,* pp. 108–109. North Holland, Amsterdam.

Van Huylenbroeck, J. M., and P. C. Debergh. 1993. Year-Round Production of Flowering *Calathea crocata:* Influence of Light and Carbon Dioxide. *Hort. Sci.* 28(9): 897–898.

Vemeulen, P. C. M., and C. W. van de Beck. 1993. Business Economic Consequences of the Environmental Targets in Relation to the Direct Energy Consumption and Carbon Dioxide Emission. Annual Report 1992–93. Glasshouse Crop Research Station, Naaldwijk, The Netherlands.

Vuylsteke, D., R. Ortiz, C. Pasberg-Gauhl, F. Gauhl, C. Gold, S. Ferris, and P. Speuer. 1993. Plantain and Banana Research at the International Institute of Tropical Agriculture. *Hort. Sci.* 28(9): 874, 970–1971.

Waggoner, P. E. 1983. Agriculture and Climate Changed by More Carbon Dioxide. In: *Changing Climate: Report of the Carbon Dioxide Assessment Committee.* pp. 383–418. National Academy Press, Washington, D.C.

Waggoner, P. E. 1984. Agriculture and Carbon Dioxide. *Am. Sci.* 72: 179–184.

Waggoner, P. E. (Ed). 1990. *Climate Change and U.S. Water Resources.* John Wiley, New York.

Waggoner, P. E. 1992. Preparing U.S. Agriculture for Global Climate Change. Task Force Report No. 119. Council for Agricultural Science and Technology, Ames, IA.

Waggoner, P. E. 1994a. How Much Land Can Ten Billion People Spare for Nature? Task Force Report No. 121. Council for Agricultural Science and Technology, Ames, IA, and the Rockefeller University, New York.

Waggoner, P. E. 1994b. Preparing U.S. Agriculture for Global Climate Change. Proc. Philadelphia Society for Promoting Agriculture 1993–94. pp. 31–45, Philadelphia, PA.

Waggoner, P. E., and R. R. Revelle. 1990. Summary. In: *Climate Change and U.S. Water Resources,* P. E. Waggoner (Ed.), pp. 447–477. John Wiley, New York.

Walsh, J. 1991. *Preserving the Options: Food Productivity and Sustainability.* Issues in Agriculture 2. Consultative Group on International Agricultural Research, Washington, D.C.

Wang F. T., S. L. Wang, Y. X. Li, and M. N. Zhong. 1991a. A Preliminary Modelling of Effects of Climate Changes on Food Production in China. In: *Climatic Variations and Change: Implications for Agriculture in the Pacific Rim.* S. Geng and C. W. Cady (Eds.), pp. 115–126. University of California, Davis.

Wang, W. C., M. P. Dudek, X. Z. Liang, and J. T. Kiehl. 1991b. Inadequacy of Effective CO_2 as a Proxy in Simulating the Greenhouse Effect of Other Radioactively Active Gases. *Nature (London)* 350: 573–577.

Warrick, R. A., and M. J. Bowden. 1981. The Changing Impacts of Droughts in the Great Plains. In: *The Great Plains Perspectives and Prospects,* M. P. Lavern, and M. E. Baker (Eds.), pp. 111–137. Center for Great Plains Studies, University of Nebraska, Lincoln.

Weenink, J. B. 1993. Environmental Policy and the Greenhouse Effect. *Vegetatio* 104/105: 357–366.

Weiss, H., M. A. Courty, W. Wetterstrom, F. Guichard, L. Senior, R. Meadow, and A. Cournow. 1993. The Georesis and Collapse of Third Millenium North Mesopotamian Civilization. *Science* 261: 995–1004.

Went, F. W. 1944. Plant Growth under Controlled Conditions, II. Thermoperiodicity in Growth and Fruiting of the Tomato. *Am. J. Bot.* 31: 135–140.

Went, F. W. 1957. The Experimental Control of Plant Growth. *Chron. Bot.* 17: 1–126.

Wetzel, R. G., and J. B. Grace. 1983. Aquatic Plant Communities. The Response of Plants to Rising Levels of Atmospheric Carbon Dioxide. In: *CO_2 and Plants,* E. R. Lemon (Ed.), pp. 223–280. AAA Selected Symposium 84. Westview Press, Boulder, CO.

Wheeler, R. M., T. W. Tibbitts, and A. H. Fitzpatrick. 1991. Carbon Dioxide Effects in Potato Growth under Different Photoperiods and Irradiance. *Crop Sci.* 31: 1209–1213.

White, G. F. 1984. Water Resource Availability, Illusion and Reality. In: *The Resourceful Earth,* J. L. Simon, and H. Kahn (Eds.), pp. 251–271. Basil Blackwell, Oxford, England.

White, M. R. 1985. *Characterization of Information Requirements for Studies of CO_2 Effects: Water Resources, Agriculture, Fisheries, Forests and Human Health.* U.S. Department Energy, DOE/ER-0236, Washington, D.C.

Wigg, D. 1994. *The Quiet Revolutionaries.* World Bank Essays. World Bank, Washington, D.C.

Wigham, D. K. 1980. Soybean. In: *Potential Productivity of Field Crops under Different Environments.* pp. 205–225. International Rice Research Institute, Manila.
Wigley, T. M. L. 1987. Relative Contributions of Different Trace Gases to the Greenhouse Effect. *Climate Monitor* 16: 14–28.
Wildavsky, A. 1992. *Introduction to the Heated Debate: Greenhouse Predictions Versus Climate Reality.* by Robert C. Balling. Pacific Research Institute for Public Policy, San Francisco, CA.
Williams, G. V. D. 1985. Estimated Bioresource Sensitivity to Climate Change in Alberta. *Climate Change* 7: 55–69.
Williams, L. E., T. M. Delong, and D. A. Phillips. 1981. Carbon and Nitrogen Limitations on Soybean Seedling Development. *Plant Physiol.* 68: 1206–1209.
Willets, D. H., and M. M. Peet. 1989. Prediting Yield Responses to Different Greenhouse CO_2 Enrichment Schemes: Cucumbers and Tomatoes. *Agr. For. Meteorol.* 44: 275–293.
Wisniewski, J., and A. E. Lugo. 1992. Natural Sinks of CO_2. *Water Air Soil Poll.* 64: Nos. 1–2.
Wittwer, S. H. 1970. Aspects of CO_2 Enrichment for Crop Production. *Trans. ASAE* 13: 240–251, 256.
Wittwer, S. H. 1978a. The Next Generation of Agricultural Research. *Science* 199: 4327.
Wittwer, S. H. 1978b. Carbon Dioxide Fertilization of Crop Plants. In: *Crop Physiology.* U. S. Gupta (Ed.), pp. 310–333. Oxford and IBH Publ., New Delhi, India.
Wittwer, S. H. 1980. Carbon Dioxide and Climate Change: An Agricultural Perspective. *Journal of Soil and Water Conservation* 35(3): 116–120.
Wittwer, S. H. 1981a. The 20 Crops That Stand Between Man and Starvation. *Farm Chem.* 144(9): 17.
Wittwer, S. H. 1981b. The Intricate Measure: An Agricultural Perspective on Research Tools for Early Detection of the Biological, Biochemical, and Environmental Sensitivities of CO_2. In: *Proceedings of the Workshop on First Detection of Carbon Dioxide Effects.* pp. 434–457. Harpers Ferry, WV, June 8–10. U. S. Department of Energy, Washington, D.C.
Wittwer, S. H. 1985. Carbon Dioxide Levels in the Biosphere: Effects on Plant Productivity. CRC *Crit. Rev. Plant Sci.* 2(3): 171–198.
Wittwer, S. H. 1986. Worldwide Status and History of CO_2 Enrichment: An Overview, In: *Carbon Dioxide Enrichment of Greenhouse Crops: Vol. 1. Status and CO_2 Sources.* H. Z. Enoch and B. A. Kimball (Eds.), pp. 3–15. CRC Press, Boca Raton, FL.
Wittwer, S. H. 1988. *The Greenhouse Effect.* Carolina Biology Reader 163. Carolina Biological Supply Company, Burlington, N.C.
Wittwer, S. H. 1990. Implications of the Greenhouse Effect on Crop Productivity. *HortScience* 25(2): 1560–1567.
Wittwer, S. H. 1992. In Praise of Carbon Dioxide. *Policy Rev.* 62: 4–9.
Wittwer, S. H. 1993. Worldwide Use of Plastics in Horticultural Production. *HortTechnology* 3(1): 6–19.
Wittwer, S. H. 1994. Impact of the Greenhouse Effect on Plant Growth and Crop Productivity. In: *Mechanisms of Plant Growth and Improved Productivity.* A. S. Basra (Ed.), pp. 199–228. Marcel Dekker, New York.
Wittwer, S. H., and L. Haseman. 1946. Soil Nitrogen and Thrips Injury in Spinach. *Science* 103: 331–332.
Wittwer, S. H., H. Strallworth, and M. J. Howell. 1948. The Value of a "Hormone" Spray for Overcoming Delayed Fruit Set and Increasing Yields of Outdoor Tomatoes. *Proc. Am. Soc. Hort. Sci.* 51: 371–380.
Wittwer, S. H., and W. Robb. 1964. Carbon Dioxide Enrichment of Greenhouse Atmospheres for Food Crop Production. *Econ. Bot.* 18: 34–56.

Wittwer, S. H., and M. J. Bukovac. 1969. Uptake of Nutrients Through Leaf Surfaces. In: *Handbuch der Planzenernahrung und Dungung.* K. Scharrer and H. Linser (Eds.), pp. 235–261. Springer-Verlag, New York.

Wittwer, S. H., and S. Honma. 1979. *Greenhouse Tomatoes, Lettuce and Cucumbers.* Michigan State University Press, East Lansing.

Wittwer, S. H., Yu Youtai, Wang Lianzheng, and Sun Han. 1987. *Feeding a Billion: Frontiers in Chinese Agriculture.* Michigan State University Press, East Lansing.

Wittwer, S. H., and N. Castilla. 1995. Protected Cultivation of Horticultural Crops—Worldwide. *Hort. Technol.* 4. 5(2): 6–24.

Wong, S. C. 1980. Effects of Elevated Partial Pressures of CO_2 on Rate of CO_2 Assimilation and Water Use Efficiency in Plants. In: *Carbon Dioxide and Climate.* G. I. Pearman (Ed), pp. 159–166. Australian Academy of Science, Canberra.

Wong, S. C. 1993. Interaction Between Elevated Atmospheric Concentration of CO_2 and Humidity on Plant Growth: Comparison Between Cotton and Radish. *Vegetatio* 104/105: 211–221.

Woodman, J. N., and C. S. Furiness. 1989. Potential Effects of Climate Change on U.S. Forests: Case Studies of California and the Southeast. In: *The Potential Effects of Global Climate Change on the United States.* Appendix D. J. B. Smith, and D. A. Tirpak (Eds.). (EPA 230-05-89-054), U.S. Environmental Protection Agency, Washington, D.C.

Woodward, F. I. 1987. Stomatal Numbers Are Sensitive to Increases in CO_2 from Pre-Industrial Levels. *Nature (London)* 617–618.

Woodward, F. I. 1993. Plant Responses to Past Concentrations of CO_2. *Vegetatio* 104/105: 145– 155.

Woodward, F. I., and F. A. Bazzaz. 1988. The Responses of Stomatal Density to CO_2 Partial Pressure. *J. Exp. Bot.* 39: 1771–1791.

Woodwell, G. M. 1970. Effects of Pollution on the Structure and Physiology of Ecosystems. *Science* 168: 429.

Woodwell, G. M. 1993. *Policy Review,* No. 63, p. 93. The Heritage Foundation, Washington, D.C.

World Commission on Environment and Development. 1987. *Our Common Future.* Oxford University Press, New York.

World Food Council. 1992. *Report to the United Nations Council Assembly.* United Nations, New York.

World Resources Institute. 1992–93. *Global Biodiversity Strategy.* The World Resources Institute, Washington, D.C.

World Resources Institute. 1994–95. *World Resources, 1994–95.* pp. 197–212. Oxford University Press, New York.

Wuebbles, D. J., and J. Edmonds. 1991. *Primer on Greenhouse Gases.* Lewis Publishers, Chelsea, MI.

Yoshida, S. 1973. Effects of CO_2 Enrichment at Different Stages of Panicle Development on Yield Components and Yield of Rice (*Oryza sativa* L.). *Soil Sci. Plant Nutr.* 19: 311–316.

Yoshida, S. 1976. Carbon Dioxide and Yield of Rice. In: *Climate and Rice.* pp. 211–221. International Rice Research Institute, Los Banos, Philippines.

Yoshida, S. 1981. *Fundamentals of Rice Crop Science.* International Rice Research Institute, Los Banos, Philippines.

Yoshida, S., and F. T. Pardo. 1976. Climatic Influence on Yield and Yield Components of Lowland Rice in Tropics. In: *Climate and Rice.* pp. 471–494. International Rice Research Institute, Manila.

Young, J. E. 1993. *Global Network. Computers in a Sustainable Society.* Worldwatch Paper 115. Worldwatch Institute, Washington, D.C.

Zak, D. R., K. S. Pregitzer, P. S. Curtis, J. A. Teeri, R. Fogel, and D. L. Randlett. 1993. Elevated Atmospheric CO_2 and Feedback Carbon and Nitrogen Cycles. *Plant Soil* 151: 105–117.

Zangerl, A. R., and F. A. Bazzaz. 1984. The Response of Plants to Elevated CO_2. II. Competitive Interactions Between Annual Plants under Varying Light and Nutrients. *Oecologia* 62: 412–417.

Ziska, L. H., A. H. Teramura, J. H. Sullivan, and A. McCoy. 1993. Influence of Ultra-violet-B (UV-B) Radiation on Photosynthetic and Growth Characteristics in Field-Grown Cassava (Manihot esculentum). Crantz. *Plant, Cell Environ.* 16: 73–79.

INDEX